SPINOFF

1989

National Aeronautics and
Space Administration
Office of Commercial Programs
Technology Utilization Division
by James J. Haggerty
August 1989

ON THE COVER:
A memento of history's most successful unmanned space exploration, the exciting 12-year odyssey of Voyagers 1 and 2 that culminated in an August 1989 Neptune encounter after earlier visits to Jupiter, Saturn and Uranus. The cover art is a false-colored Voyager 1 image of Saturn's moon Titan showing layers of haze in Titan's atmosphere.

For sale by the
Superintendent of Documents,
U.S. Government Printing Office
Washington, D.C. 20402

Foreword

President George Bush with NASA Administrator Richard H. Truly (center) and Deputy Administrator James R. Thompson.

Speaking on the 20th anniversary of the first Apollo lunar landing, President Bush set the direction for the United States' civil space program well into the 21st century. The President proposed a long range, continuing commitment to a "sustained program of manned exploration of the solar system—and yes—the permanent settlement of space."

Under the President's program, the nation would complete Space Station *Freedom* within the next decade; return to the Moon early in the next century... this time to stay... and after that, send a manned expedition to explore Mars.

NASA is paving the way toward extending human presence beyond Earth orbit with a broad and comprehensive research and technology development program for the remaining years of this century. We expect this program to yield countless benefits, including a harvest of technological spin-offs and applications to benefit life on Earth.

Together with our partners in Europe, Japan and Canada, we are moving ahead on Space Station Freedom. We have successfully restored the Space Shuttle to operational status, we are gradually increasing flight frequency, and we are developing new vehicles and systems that will significantly enhance our capabilities in space transportation.

We have embarked on a golden era of space science. In this year and through 1993, NASA plans 36 important space science missions the highest rate in the history of the U.S. space program. Much of our space science and applications effort will center attention on Planet Earth in a quest for answers as to how and why our planet is changing and what we—all the people of the Earth—can do and must do to cope with future global change.

In the interest of national technological competitiveness, we are forging a working partnership with U.S. industrial firms to exploit the commercial space potential, and we are seeking to broaden the range of opportunities for promising commercial space ventures.

We are also conducting a program of advanced aeronautical research that offers practical benefits in safer, better performing, more efficient, more environmentally acceptable aircraft. In the process, we are developing many technologies that have applicability to space as well as to aeronautics.

Finally, we are laying a foundation for tomorrow with a developmental effort focused on emerging, innovative technologies that can make possible the manned lunar/planetary missions and robotic explorations contemplated.

NASA is poised for the challenges ahead. Successful execution of these programs will position NASA well for the next giant-step and, with the support of the American people, we are confident we can surmount them. As President Bush reminded us:

"To this day, the only footprints on the Moon are American footprints. The only flag on the Moon is the American flag. And the know-how that accomplished these feats is American know-how. What Americans dream, Americans can do."

Richard H. Truly
Administrator
National Aeronautics and Space Administration

Introduction

For more than three decades, NASA has been developing advanced technology to meet its aeronautical and space goals. Much of the hardware developed, having served its purpose, is no longer extant—but the technology remains.

Technology is simply knowledge, technical know-how, and like other forms of knowledge, it is readily transferable. Once developed, a technology can be reapplied to uses different—and often remote—from the original application. Thus, the great storehouse of technology that NASA has built up constitutes a valuable national resource, a bank of know-how available for secondary application, or "spinoff."

Tens of thousands of spinoff products and processes have emerged from the reapplication of technology developed for NASA mainline programs, each contributing some measure of benefit to the national economy, productivity or lifestyle. These spinoffs span so broad a range of public needs and conveniences that it is difficult to find an area of everyday life they have not pervaded. Collectively, they represent a substantial dividend on the national investment in aerospace research.

The technology bank is still growing daily and the Congress, recognizing its great potential, has charged NASA with stimulating the widest possible use of this valuable resource in the national interest. NASA's instrument of that purpose is the Technology Utilization Program, which seeks to broaden and accelerate the technology transfer process. Its intent is to spur expanded national benefit, in terms on new products and new jobs, by facilitating technology transfer—providing a channel linking the technology and those who might be able to put it to advantageous use.

This publication is an implement of the Technology Utilization Program intended to heighten awareness among potential users of the technology available for transfer and the economic and social benefit that might be realized by secondary applications.

Spinoff 1989 is organized in three sections:

Section 1 outlines NASA's mainline effort, the major programs that generate new technology and, therefore, replenish and expand the bank of knowledge available for transfer.

Section 2, the focal point of this volume, contains a representative sampling of spinoff products that resulted from secondary application of technology originally developed to meet mainline goals.

Section 3 describes the various mechanisms NASA employs to stimulate technology and lists, in an appendix, contact sources for further information about the Technology Utilization Program.

James T. Rose
Assistant Administrator for
Commercial Programs
National Aeronautics and
Space Administration

Contents

AEROSPACE AIMS

An illustrated summary of NASA's major aeronautical and space programs, their goals and directions, their contributions to American scientific and technological growth, and their potential for practical benefit

Probing the Universe

Our nearest planetary neighbor, cloud-shrouded Venus, approximates Earth in age, size, mass, density and orbital distance from the Sun. Yet in water content, atmosphere, temperature, pressure and many other aspects, the "twin" planets are markedly different.

Why, scientists want to know, did two similar planets evolve in such dissimilar fashion. Their interest goes well beyond academic curiosity; they feel that the answers hold clues to greater understanding of the many factors that influence Earth's complex environment.

NASA's space science and applications program seeks greater knowledge of the universe and expanded Earth benefits through application of space technology

Although Venus has been observed first hand by more than a score of U.S. and Soviet spacecraft and is the most visited of all the planets, scientists have few of the answers they seek. But they await with mounting excitement the 1990 arrival of the next Venus visitor, the radar-imaging Magellan spacecraft launched on May 4, 1989. Magellan is expected to amplify manyfold the accumulated information about Venus, acquire the best-yet pictures of the planet's hidden "face," and unveil many of the secrets of the Venusian past.

Since the planet's permanent cloud cover bars conventional photography of Venus, scientists have employed—for more than a quarter of a century—radar systems to penetrate the clouds and provide data for pictures derived from computer processed radar reflections. Earth-based radar imaging has proved valuable, but it is limited in that only a fraction of the planet can be covered. Spacecraft—notably NASA's Pioneer Venus and the USSR's Venera 15 and 16—provided global data to answer many questions about Venus' atmosphere and large-scale surface features, but they have not had the resolution—a measure of the smallest objects than can be seen on radar maps—to allow scientists to piece together a solid geologic history of Venus.

Magellan will map nearly the entire planet with 10 times the detail of the best previous spacecraft images and 100 times the resolution of its predecessor, Pioneer Venus. Its images will reveal, for the first time, such small-scale features as hills and valleys, craters and lava flows, keys to Venus' geology and planetary history. The images will provide information about the extent to which Venus' surface has been influenced by volcanoes, plate tectonics, meteorite craters, water and wind erosion, and other processes that determine a planet's history and shape its face.

Although Magellan is a large spacecraft, weighing almost four tons on Earth, it carries only one science instrument, a radar sensor known as a Synthetic Aperture Radar (SAR). The SAR will perform three important functions: collecting data for surface imaging; measuring surface height variations as small as 100 feet to construct a topographic profile of Venus; and measuring the natural thermal emissions from the planet to determine surface temperature variations. Additional information will be provided by Earth-based tracking of Magellan's orbit around Venus, which will detect slight changes in the planet's gravitational field and thereby provide clues to the makeup of the planet's interior.

The Magellan program is managed for NASA by Jet Propulsion Laboratory. Martin Marietta Astronautics Group is prime contractor for the spacecraft and Hughes Aircraft developed the SAR.

Magellan will complete a 158-million-mile roundabout journey to Venus in August 1990, at which time the spacecraft's solid rocket motor will fire to place Magellan in orbit around the planet. Since Venus has a very long "day," equal to 243 Earth days, the primary mapping mission will take 243 days because it will take that long for most of the planet's surface to pass beneath the orbiting radar's gaze. During that time, Magellan will map 90 percent of the surface and will acquire more digital imaging data than all previous U.S. planetary missions combined. The program will contribute important new volumes to the science of comparative planetology—relating phenomena on one planet to conditions on another—and provide a wealth of information on the intriguing matter of why Earth's neighbor planet Venus is a twin, yet a stranger.

This artist's rendering shows Magellan in elliptical orbit around Venus and illustrates the mapping and data transmission phases of the mission. For 37 minutes on each orbit, when Magellan is close to the planet's surface, the radar system will map a 16-mile-wide swath; as the spacecraft reaches the most distant point of its orbit, it will transmit its data to Earth stations.

Shown at launch on May 4, 1989 from the Shuttle Orbiter Atlantis, the Magellan spacecraft is affixed to the Boeing-built Inertial Upper Stage booster that sent it on a 15-month journey to Venus for the most comprehensive surface mapping of the planet ever undertaken.

Voyager at Neptune

At right is an artist's concept of the historic moment on August 24, 1989, when the Voyager 2 spacecraft reached the closest point in its encounter with the giant gaseous planet Neptune, a world never before visited by spacecraft. The smaller celestial body in the photo is Neptune's moon Triton, object of Voyager's imaging scrutiny five hours later.

At the time of the close encounter, Voyager 2 was 12 years out of home port Earth and had traveled more than four billion miles on its grand tour of the outer planets Jupiter, Saturn, Uranus and Neptune. Voyager 2 passed within 3,000 miles of Neptune, which was then some 2.8 billion miles from Earth. Voyager's radio signals, traveling at the speed of light, took four hours and six minutes to reach Earth.

Months before the closest approach, Voyager 2 was returning Neptune images in which light and dark spots kept appearing and disappearing. This was evidence of an unexpectedly dynamic atmosphere—unexpected because solar energy drives planetary atmospheres and distant Neptune receives only one-tenth of one percent as much solar energy as does Earth.

The photo at top right shows views taken April 26, 1989, when the spacecraft was still 109 million miles from the planet. The left image shows a great dark spot just above the equator; this was an early view of an exciting discovery, an enormous cloud system bigger than the whole planet Mars, an indicator of chemical reactions in the atmosphere that produce dark, probably carbon-rich material.

In the middle right photo are five views of Neptune, each taken with a different filter on May 24 at a distance of 83 million miles. The top three show clear views of the massive black spot and the ring around the South Pole. The lower right photo was taken on June 22 at 57 million miles. The image on the left shows Neptune as it would appear to the unaided eye

if the eye could get close enough; the other image is a computer-enhanced view that better outlines the dark spot and the polar "collar."

On July 5, Voyager's sensors made a major discovery, evidence of a third moon of Neptune.

The close encounter period began on August 21 and continued for nine days; during that time Voyager took closeup views of Neptune, its arc-like rings and the moon Triton, pictures of inestimable value to science because they are the first clear views ever obtained of the distant planet. Neptune is not visible to the unaided eye and even to large Earth-based telescopes it shows up as a small, greenish disc in which no surface detail is discernible.

The Voyager mission is managed for NASA by the California Institute of Technology's Jet Propulsion Laboratory, which also manages

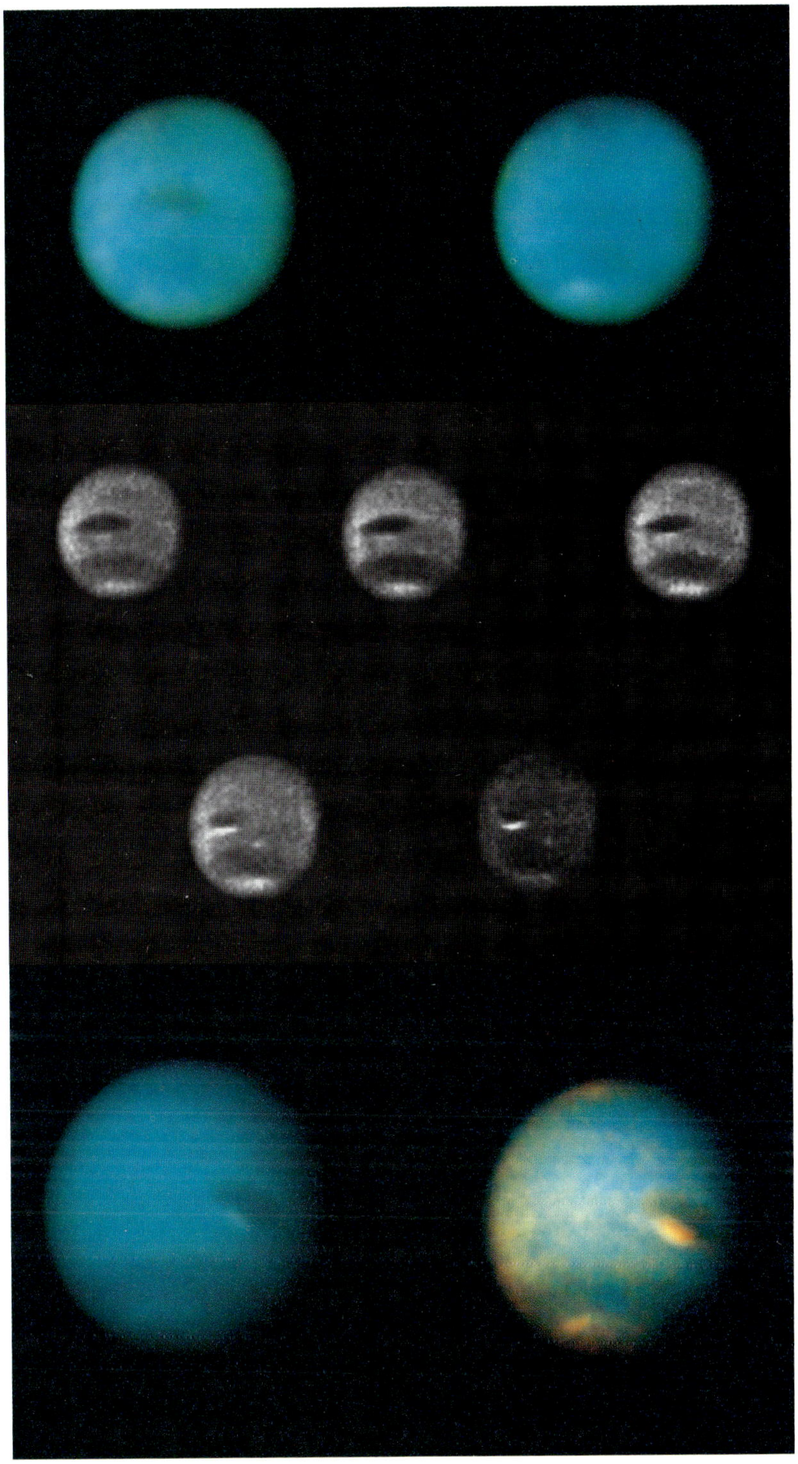

NASA's Deep Space Network (DSN), a system of radio telescope complexes that provides the communications link with Voyager 2. The DSN stations are strategically located near Barstow, California; Madrid, Spain; and Canberra, Australia so that, as Earth rotates, at least one station is in direct "line-of-sight" with Voyager.

To track the faint signals from the spacecraft's transmitter, which has about the same wattage as a refrigerator light bulb, the DSN's supersensitive antennas have been considerably upgraded. To increase data return, the nine DSN antennas are being supplemented by other tracking facilities: the Very Large Array (27 antennas) near Socorro, New Mexico, operated for the National Science Foundation by Associated Universities, Inc.; the Parkes Radio Observatory near Canberra, operated by the Australian Commonwealth Scientific and Industrial Research Organization; and Usuda Observatory, Honshu, Japan, operated by the Institute of Space and Astronautical Science of Japan.

The Voyager science team includes some 150 scientists from the U.S., Canada, the United Kingdom, France, West Germany and Italy who work on Voyager's 11 scientific investigations.

Voyager 2 departed Earth on its grand tour August 20, 1977, followed 16 days later by Voyager 1. Both spacecraft visited Jupiter and Saturn, Voyager 2 went on to Uranus and Neptune. Collectively, the two spacecraft returned thousands of images of the outer planets, 40 of their moons and several unique systems of planetary rings, and they provided additional data on planetary/interplanetary magnetic fields and ultraviolet sources among the stars.

After its Saturn encounter in 1980, Voyager 1 headed out of the solar system on a trajectory that takes it above the ecliptic, the plane in which Earth and most of the planets orbit the Sun. By the next century, it may cross the as yet undefined heliopause, the boundary between the Sun's magnetic influence and interstellar space. Voyager 2 will similarly head out of the solar system, but on a flight path that will take it below the ecliptic. The Voyagers are expected to continue reporting data until about 2020 when their nuclear power supply will be depleted and they will cruise on silently through interstellar space.

Solar System Exploration

This year marks a renaissance of U.S. planetary exploration, with Magellan en route to Venus, Voyager 2 concluding its encounter with Neptune, and a third major spacecraft—the Jupiter-orbiting observatory Galileo—poised for Shuttle launch in late 1989.

Galileo (right center) will provide the first direct sampling of Jupiter's atmosphere and the first extended observations of the planet, its moons and its intense magnetospheric environment. It will allow the first closeup look at asteroids, along with extensive observations of Venus and the Earth-Moon system.

These latter investigations come as bonuses from a change in Galileo's flight path necessitated by cancellation of the orginally planned Centaur upper stage booster and its replacement by a less powerful Inertial Upper Stage (IUS). After release in Earth orbit from the Shuttle Orbiter, Galileo will be boosted to Earth-escape velocity by the Boeing-built IUS. The reduced launch energy available from the IUS rules out direct boost to Jupiter; instead, the mission will employ the gravity assist technique in which planetary gravity fields are used as "slingshots" to accelerate the spacecraft and alter its trajectory. Thus, Galileo will fly a complicated path that will take it past Venus—in early 1990—for a gravity assist, then on a 10-month trip back to Earth for another gravity assist (December 1990), then on a two-year loop that will bring it back to Earth once more for a final gravity assist and a push into Jupiter trajectory (December 1992). This flight path will stretch Galileo's en route time to Jupiter to almost six years.

In 1992, between the first and second Earth gravity assist passes, Galileo will fly past a 10-mile-diameter main belt asteroid known as Gaspra; it will pass within 620 miles of the asteriod and take hundreds of pictures. In 1993, while en route to Jupiter, the spacecraft will fly by Ida, an asteroid about twice as large as Gaspra, and repeat the imaging process. Galileo's instruments will also measure the surface composition and roughness, optical and thermal properties, rotation rate and mass of the two asteroids.

Galileo's Jupiter exploration has three general and equal objectives: investigation of the structure and physical dynamics of the planet's complex atmosphere; measurement of chemical composition and physical state of the Jovian moons; and determination of the composition of Jupiter's atmosphere. The latter assignment will be accomplished by an instrumented probe that will be released from the main spacecraft in July 1995 to make an independent five-month trip to the planet. The probe will then descend by parachute into the Jovian atmosphere, sending data to the main spacecraft for up to 75 minutes.

The main spacecraft, or orbiter, will make its closest approach (620 miles) to Jupiter on December 7, 1995 and swing into orbit around the giant planet. Galileo's primary mission is planned to last 22 months, during which the spacecraft will make 10 orbits and make a close flyby of one Jovian moon on each orbit, imaging the moons with resolutions 20 to 100 times better than any previous pictures.

Galileo is a cooperative project with the Federal Republic of Germany. It is managed for NASA by Jet Propulsion Laboratory (JPL), which designed and built the orbiter. Ames Research Center has responsibility for the probe, which was built by Hughes Aircraft.

A solar system exploration program of importance is Ulysses, a cooperative endeavor of NASA and the European Space Agency (ESA), with JPL as NASA's project manager. Ulysses (top right) will embark in 1990 on a multiyear mission that will take it out of the ecliptic, the imaginary plane through the solar system that

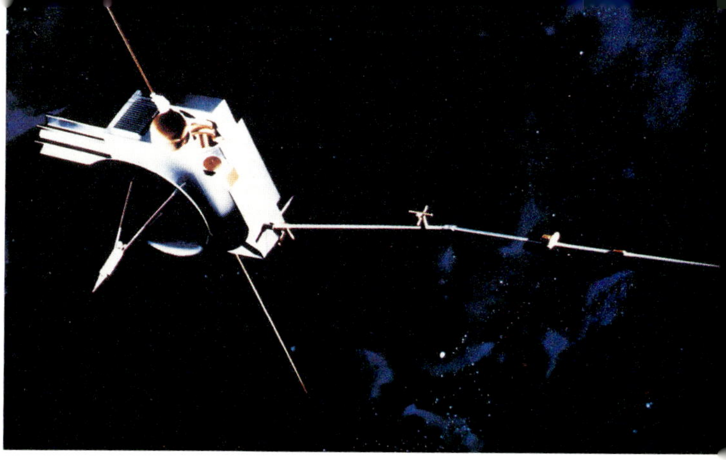

approximates an extension of the Sun's equator. Since all spacecraft have operated close to that plane, the space above and below the ecliptic is still unexplored and of great scientific interest. Flying a path that will take it eventually over the Sun's south pole, Ulysses will report—from the fresh perspective of a never-before penetrated region of the solar heliosphere—on such features as the solar wind, solar and galactic radiation, cosmic dust and solar/interplanetary magnetic fields.

Planned for launch in 1992, the Mars Observer (right) will make a two-year global survey of Mars from orbit, providing remotely sensed data of the planet's surface and atmosphere with the highest resolution yet attained and collecting new information about two general areas: geoscience and climatology. JPL manages the program; General Electric Astro Space is developing the spacecraft. Mars Observer is a multinational project with participation by Australia, the Federal Republic of Germany, France, the United Kingdom and the Soviet Union.

The Mars Observer will provide an "extensive inheritance"—a common Planetary Observer bus and several instruments—to a planned but not yet authorized Lunar Observer, designed to carry out a one-year orbital survey of the Moon. Tentatively planned for the 1990s, it will conduct a global lunar mapping mission, measure the Moon's mineral and elemental composition, assess its resources, measure surface topography and magnetic gravitational fields.

Another major solar system exploration mission planned is the JPL-managed Comet Rendezvous Asteroid Flyby (CRAF)/Cassini mission, a cooperative project with ESA. CRAF/Cassini is a multi-objective, multiyear program combining into a single initiative what were originally two separate programs. Tentatively planned for launch in 1995, CRAF will fly by an asteroid and then make a rendezvous with the Comet Kopff for closeup study of its nucleus, dust and atmosphere over a period of more than three years. The Cassini spacecraft, intended for 1996 launch, will make a flyby of an asteroid, then employ a Jupiter gravity assist toward Saturn. It will spend four years in orbit around Saturn conducting a detailed exploration of the Saturnian system. Building on data supplied by NASA's Pioneers and Voyagers, the CRAF/Cassini program will provide unprecedented information on the origin and evolution of the solar system and shed light on how the building blocks for life are formed in the universe.

Great Observatories

Earth's atmosphere admits visible light and radio waves emitted by celestial bodies, but bars most other forms of energy from reaching Earth's surface. Even the finest Earth-based observatories, therefore, are limited in observational capability because they view only in visible light and the light they detect is distorted by the layer of atmosphere.

The advent of orbiting spacecraft opened up a whole new astronomical vista. Satellite-based telescopes operating above the atmosphere get an undistorted view of the universe, and they can observe in portions of the electromagnetic spectrum other than visible light—ultraviolet, infrared, x-rays and gamma rays, for example. The latter capability is particularly important to astronomical science, because each segment of the spectrum offers a different set of clues to the origin and evolution of the universe.

NASA has flown a number of orbital observatories, each helping to explain different processes behind astronomical phenomena. But, until now, satellite-based telescopes have also suffered restrictions, imposed by their relatively small size and limited spectral capability.

The decade of the 1990s promises the most dramatic advances in the history of astronomy as NASA introduces to service four "Great Observatories," each operating in a special segment of the electromagnetic spectrum so that collectively they can view the full range of phenomena in the universe. They will enable scientists to study the cosmic happenings of the early years of the universe and produce a comprehensive picture of the cosmos that no single observatory could provide.

The first of the Great Observatories is the Hubble Space Telescope (HST), scheduled for launch in the first quarter of 1990. Described as the most important scientific instrument ever designed for use in orbit, the HST will literally look back in time some 14 billion years, observing the universe at a time when galaxies were being formed.

The HST will view in the visible light and ultraviolet wavelengths. It will enormously expand the observable universe, return images with extraordinary clarity and detect very dim objects that have never been seen. The observatory's size is evident above, where the HST is shown, covered with its thermal protective coating, being moved to a vacuum chamber for testing. Developed by Lockheed Missiles & Space Company, the spacecraft is 43 feet long and weighs 12 1/2 tons. Perkin-Elmer Corporation is prime contractor for the optical assembly. Marshall Space Flight Center managed the development program; Goddard Space Flight Center will control the telescope and process its data when the HST is in orbit. The European Space Agency (ESA) furnished the power-generating solar arrays and one of the system's five major instruments.

Planned for launch in midyear 1990 is the second of the Great Observatories, the Gamma Ray Observatory (upper right), which will investigate gamma radiation, the most energetic of all forms of radiation, and its violent sources—pulsars, quasars, black holes and other objects that have not been identified in any other wavelengths.

Managed by Goddard Space Flight Center, the Gamma Ray Observatory is a joint development of the U.S., the Federal Republic of Germany, the Netherlands, the United Kingdom and ESA; TRW Inc. is the U.S. prime contractor.

Third of the Great Observatories is the Advanced X-ray Astrophysics Facility (AXAF), a 10-ton observatory targeted for launch in the mid-1990s. It is designed to obtain high resolution x-ray imagery and study such subjects as stellar black holes, the contribution of hot gas to the mass of the universe, clusters and superclusters of galaxies, the existence of "dark matter" in the universe, the age and ultimate fate of the universe. AXAF (below right) will have instruments at least 100 times more sensitive than those of the High Energy Astronomical Observatory-2 that provided the most comprehensive data yet acquired about celestial X-ray sources. AXAF is managed by Marshall Space Flight Center, with TRW Inc. as major contractor. Foreign participation includes the Federal Republic of Germany, The Netherlands and the United Kingdom.

Not yet approved for development but planned for launch in the late years of this century is the fourth of the Great Observatories, the Space Infrared Telescope Facility (SIRTF). SIRTF is envisioned as a follow-on to the Infrared Astronomical Satellite (IRAS) that mapped some 100,000 infrared sources in 1983. SIRTF will study areas of high interest identified by IRAS, including cosmic births of galaxies and stars, powerful infrared-emitting sources at the edge of the universe, and the "missing mass," the 90 percent of the matter in the universe that is invisible and thought to exist because of the gravitational forces apparently exerted on stars and galaxies. Ames Research Center is SIRTF project manager.

Astronomy and Astrophysics

In addition to the Great Observatories, NASA's astronomy and astrophysics program embraces a number of other spacecraft whose observations of celestial phenomena will complement and reinforce the findings of the large observatories. One such is the Cosmic Background Explorer (COBE), scheduled for launch in November 1989 to seek evidence supporting the Big Bang theory of the origin of the universe.

The Big Bang theory holds that the universe began some 15 billion years ago with a monumental explosion, which triggered a uniform expansion of the universe that has continued ever since. The best evidence of the primeval explosion is the uniform background radiation discovered in 1964 by radio telescope scientists of Bell Telephone Laboratories; they detected signals coming from space, not just from one object but equally from all directions. This phenomenon cannot be explained except as a remnant of the Big Bang; it offers clues to the processes that shaped the expansion of the universe.

Until now, this radiation has been very difficult to study because it is faint and easily absorbed by Earth's atmosphere. But COBE (right) will provide enormous improvements in instrument sensitivity and data accuracy. Its three instruments will map the cosmic background, determine the detailed spectrum of the background radiation and search for the earliest-formed galaxies. COBE is managed by Goddard Space Flight Center.

Also managed by Goddard is ROSAT (for Roentgensatellite), a joint NASA/United Kingdom/Federal Republic of Germany development designed as a stepping stone toward the Advanced X-ray Astrophysics Facility (see previous page). Targeted for launch in the spring of 1990, ROSAT (top right) has an x-ray telescope and imaging system that will conduct a sweeping survey of x-ray sources and make dedicated observations of specific sources, allowing astronomers to study in greater detail many of the phenomena discovered, but not thoroughly investigated, by earlier x-ray satellites.

Planned for launch in 1991 is the Extreme Ultraviolet Explorer (EUVE) shown at middle right; the designation refers to a wavelength band between the ultraviolet and x-ray ranges that has never been explored. EUVE's four sensitive instruments will conduct an all-sky survey in that band of the spectrum. The project is managed by Jet Propulsion Laboratory. Major contractors include Fairchild Space Company, General Electric Company and McDonnell Douglas Astronautics Company.

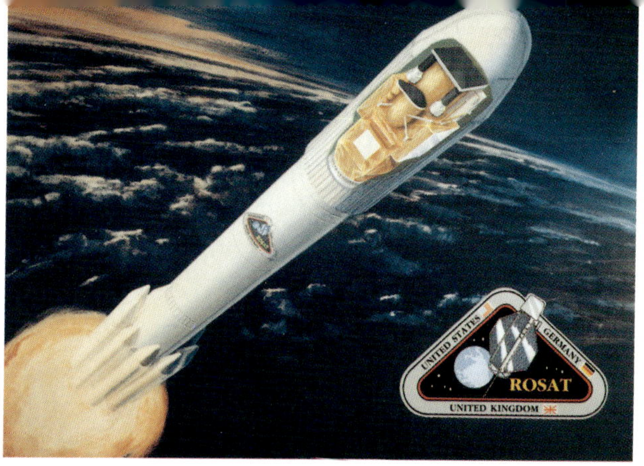

A NASA Shuttle-based astronomical system known as the Astro Observatory (bottom right) will complement the investigations of the Hubble Space Telescope by studying quasars, galaxies and active nuclei in the far ultraviolet range. Additionally, the system is capable of measuring x-ray emissions from stars, galaxies and supernova remnants. Its full complement of instruments includes three ultraviolet telescopes aligned to each other on a single pointing system and an x-ray telescope with a separate pointing system. Mounted on Spacelab pallets, these instruments can perform independent or simultaneous observations of selected targets and conduct as many as 300 observations during a 10-day Shuttle mission.

Astro 1, which will be configured without the x-ray telescope, is planned for flight in the spring of 1990. Astro 2, targeted for 1992 service, will carry the full four-instrument complement. Marshall Space Flight Center is project manager.

Planned for development but not yet approved is the Stratospheric Observatory for Infrared Astronomy (SOFIA), an airborne observatory that will provide capabilities complementary to those of the Space Infrared Telescope Facility (see previous page). The SOFIA system includes a three-meter telescope mounted in a Boeing 747 transport for observations at altitudes above 40,000 feet in infrared wavelengths inaccessible from the ground. SOFIA is a cooperative project with the Federal Republic of Germany; Ames Research Center leads the project management for NASA.

Solar-Terrestrial Research

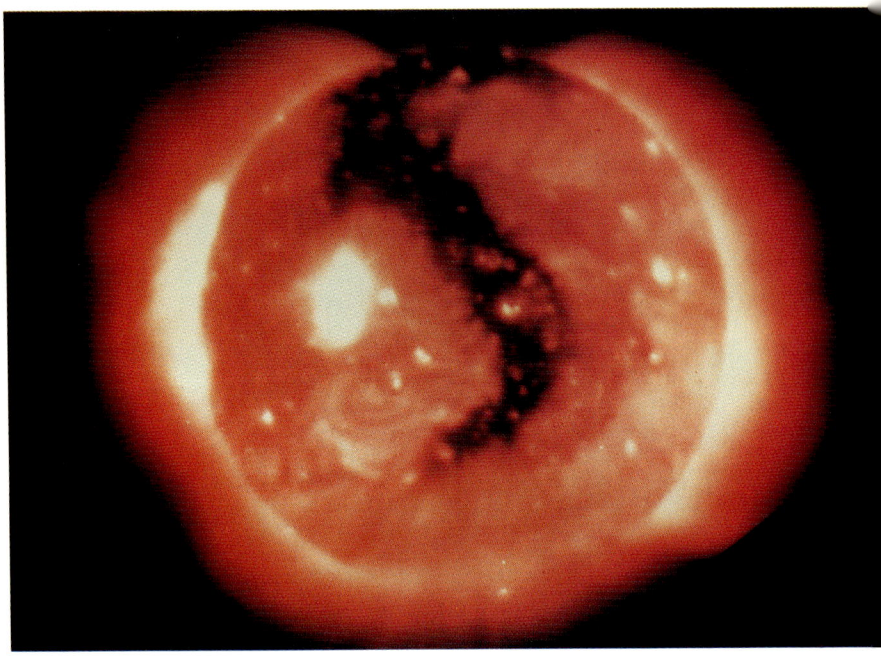

The Sun is a variable star, meaning that its activity varies over time. The changing degree of activity is due to two factors: its rotation and its convection processes that transport hot gas from the solar interior to the surface. The interaction of these two motions—rotation and convection—generates powerful magnetic fields and influences the cyclic activity level demonstrated by the ebb and flow of sunspots and solar flares.

Additionally, activity in the Sun's corona (right)—the white "halo" of gas seen during eclipses—causes ejection at very high velocities of a hot gas, or plasma, known as the solar wind, which courses through the solar system transferring energy to Earth and all the other planets; its interaction with Earth's magnetic field causes a whole range of effects, such as the aurorae (far right) magnetic storms, disruption of radio communications and power surges in transmission lines.

An important facet of NASA's space science program is solar-terrestrial research, which has underlying practical implications for Earth's climate, weather and other life-influencing factors. Solar-terrestrial research embraces the study of the Sun as a variable star, the origin and transmission of the solar wind, the interaction of the solar wind with Earth's magnetosphere, and the subsequent time-varying effects in Earth's lower atmosphere. Over the years, NASA has employed a number of spacecraft to study the physical processes that link Earth and the Sun. In the 1990s, this area of science will be the subject of an intensified, dedicated multinational collaborative effort known as the International Solar-Terrestrial Physics (ISTP) program.

ISTP will be conducted jointly by NASA, the European Space Agency (ESA) and the Japanese Institute of Space and Astronautical Science (ISAS) with possible participation by the Soviet Union. The program involves employment of several spacecraft in different orbits to make simultaneous and sequential measurements intended to establish cause and effect relationships among solar-terrestrial phenomena.

The "core" ISTP program—which will be augmented by the findings of other spacecraft conducting related and complementary research—includes five projects, each feeding information to the others, to be conducted over a multiyear period:

- *Geotail*, a cooperative NASA/ISAS effort with ISAS providing the spacecraft and both agencies contributing plasma physics instrumentation for a thorough investigation of Earth's magnetosphere and the magnetotail, the part of the magnetosphere that extends away from the Sun, something like the tail of a comet. Geotail is targeted for mid-1992 launch.
- *Polar and Wind*, two NASA spacecraft with instrumentation supplied by U.S., Japa-

nese and European investigators, intended for study of solar wind properties and processes in the magnetosphere. Planned for late 1992 launch, Wind will operate initially in a highly elliptical orbit that will take it more than a million miles from Earth. Polar, targeted for flight in 1993, will operate in an eccentric polar orbit. General Electric Astro Space is prime contractor for both spacecraft.

• *Cluster* consists of four identically instrumented spacecraft, being developed by ESA with instruments provided by both NASA and ESA, to investigate plasma (hot gas) interactions, explore the boundaries of Earth's magnetosphere and study how mass and energy are transferred across boundaries. Cluster is planned for service in the mid-1990s.

• *SOHO* (Solar and Heliospheric Observatory), being developed by ESA with shared NASA/ESA instrumentation, is designed to study the structure of the solar interior, the dynamics of the corona and the characteristics of the solar wind from an orbit between Earth and the Sun. It is planned for launch in the mid-1990s.

Goddard Space Flight Center is NASA project manager for all ISTP missions.

Earth Science and Applications

In the three decades of the Space Age, man has learned more about Earth and its environment than in all prior years. This knowledge, however, has been acquired piecemeal through separate investigations of Earth's components, such as the planet's interior, its crust, biosphere, oceans and ice cover, the atmosphere and the ionosphere.

Because all these components are interlinked, it is necessary now to study Earth as a single unified system, in order to develop a complete understanding of Earth's environment and ultimately to be able to predict global change caused by natural events or by human actions.

NASA is working toward that goal independently and in concert with its international space partners. To take advantage of new technologies that make possible advanced studies of Earth, NASA has proposed a comprehensive Earth Observing System (EOS) (right), a long range program involving multiple separate satellites, instrumented orbiting platforms, Space Shuttle missions and ground-based research. The program would include participation by the scientific organizations of other countries, in particular the European Space Agency and Japan.

NASA is further proposing a complementary Earth science program to be carried out by small-to-moderate size Explorer Class satellites with advanced instrumentation for studies of rainfall, magnetic fields and other individual components of the Earth system; instrument mapper satellites to provide high resolution detail of Earth's surface; and other spacecraft for studies of space plasma, oceanography and land processes.

A major step in global research will be accomplished by a spacecraft already in advanced development, the Upper Atmosphere Research Satellite (UARS) (top center). Being developed by General Electric Astro Space under Goddard Space Flight Center management, UARS will provide the first global measurements of the chemistry, dynamics and energetics of the upper atmosphere over a three-year span; launch is planned for the fourth quarter of 1991. Among major objectives are understanding of the mechanisms that control the structure and variability of the upper atmosphere and the role of the upper atmosphere in climate and climatic changes.

Another important project, one that is both an Earth science effort and a direct benefit application, is the Ocean Topography Experiment, also known as TOPEX/Poseidon. A joint program of NASA and the French space agency CNES, TOPEX/Poseidon is an ocean observation satellite designed to make highly accurate measurements of sea surface elevations over entire ocean basins for several years. Integrated with subsurface measurements, this information will be used in models to determine ocean circulation and its variability. TOPEX/Poseidon (top right) will significantly expand knowledge of ocean dynamics and it will also establish an informational base for practical applications, such as weather and climate prediction, coastal storm warning, maritime safety, ship design and routing, and food production from ocean sources. Jet Propulsion Laboratory (JPL) is project manager and Fairchild Space Company is developing the satellite.

A related project involves orbital flight of an advanced instrument known as the NASA Scatterometer, or NSCAT, to seek new understanding of ocean/atmosphere interaction and the relationship of that interaction to climate changes. The NSCAT is a microwave radar that enables calculation of wind speeds and directions by measurement of reflections from waves on the ocean surface. NSCAT data, together with surface elevation data from TOPEX/Poseidon, will give oceanographers their first ever view of the oceans' primary driving force (winds) and the response to that force (surface elevations), perhaps permitting a fundamental breakthrough in man's understanding of how the oceans work as a dynamic system. NSCAT will be flown in the mid-1990s aboard a satellite designated ADEOS (Advanced Earth Observation System) in a joint NASA/NASDA (Japanese Space Agency) project. JPL is NASA's project manager.

NASA will also employ Spacelab pallets to accommodate instrument payloads for Shuttle-based solar and atmospheric studies. Known as ATLAS (for Atmospheric Laboratory for Applications and Science), the project will investigate such areas as long-term changes in the total energy radiated by the Sun and the global distribution of key molecular species in the middle atmosphere. ATLAS is a cooperative program with instruments and experiments being supplied by Belgium, France, Japan and West Germany. The first ATLAS flight is scheduled for the spring of 1991, the second a year later; additional flights are planned but not approved.

NASA's major effort in space communications research is the Advanced Communications Technology Satellite (ACTS), being developed by General Electric Astro Space under the management of Lewis Research Center. Targeted for launch in 1992, ACTS

(above) represents an effort to maintain U.S. leadership in the world commercial communications satellite market and to develop communications technology that will enhance future space missions. ACTS incorporates several revolutionary changes designed to make more effective use of available frequencies, to increase the message handling capacity of the individual satellite, and to provide a capability for high volume communications relay to small Earth terminals.

NASA is also cooperating with U.S. industry in an effort to accelerate the introduction of satellite-relayed two-way voice and data communications for trucks, cars, boats and airplanes in the Mobil Satellite program. The first generation system, to be launched by NASA, is being financed and built by industry and will be operated by a U.S. corporation. Major contractors include Teledyne Brown and Ball Aerospace; Jet Propulsion Laboratory is project manager for NASA.

Toward Future Flight

Initial results of a new NASA research program suggest that rain—specifically heavy rain—may have more of an influence on airplane performance than is generally believed. And it may be important for pilots to consider the rain factor when they are flying through severe storms, because indications are that heavy rain can momentarily blur an airfoil shape, hence change the airflow and cause loss of airplane performance.

Investigation of heavy rain effect on aircraft highlights selected examples of NASA aeronautical research, which is providing technology for tomorrow's aircraft

These are the conclusions of researchers at Langley Research Center, which is studying the matter with a novel apparatus that simulates heavy rain effect on an airfoil moving at takeoff or landing speed.

Heavy rain research is part of a broader NASA/Federal Aviation Administration airborne wind shear detection and avoidance program begun in 1986. Wind shear is a sudden shift in wind velocity and direction often associated with severe storms. The most violent form of wind shear is the microburst, an intense downdraft. When a pilot encounters a microburst at low altitude on approach or takeoff, he has little time to react correctly to maintain the desired flight path. Between 1964 and 1985, there were at least 26 accidents, involving some 500 fatalities and more than 200 injuries, wherein wind shear was the direct cause or a contributing factor.

Understanding heavy rain—which by definition is high intensity, short duration rainfall—is important to wind shear investigations. Modern aircraft wings rely on a smooth, uninterrupted flow of air for maximum performance, but very heavy rain may disrupt the airflow and reduce lift. In cases where greater lift than usual is required—as in a low altitude wind shear encounter—heavy rain may reduce wing performance when it is most needed.

For most of the era of flight, little attention was paid to the influence of rain on airfoil performance because it was commonly thought to be insignificant. Wind shear researchers were the first to ask questions about the effect of heavy rain a decade ago.

In 1980, James Luers and Patrick Haines of the University of Dayton (Ohio) Research Institute, working under a grant from

NASA's Goddard-Wallops Flight Facility, performed theoretical calculations based on their hypothesis that aircraft performance is reduced during heavy rain.

Langley Research Center followed up in the mid-1980s with a series of small scale tests in a wind tunnel equipped with a spray system to simulate moderate to extremely heavy rainfall on an instrumented wing model. Those tests indicated a definite effect—lift losses of more than 20 percent—at extremely high rain rates.

The results of the small scale tests prompted NASA to seek verification at larger scale and with more accurately simulated flight conditions. That led to the current series of tests in which a full scale wing section mounted on a tubular steel carriage is accelerated to airplane takeoff speed, then subjected to a deluge of heavy rain.

(Continued)

At Langley Research Center, NASA is using this track and carriage facility to investigate the effect of heavy rain on aircraft wings.

Toward Future Flight
(Continued)

Known as the Aircraft Landing Dynamics Facility, the track and carriage apparatus Langley Research Center is using for heavy rain research is normally employed to evaluate landing gear systems and components. It offers safety, economy, control and versatility advantages over flight testing.

Use of the facility dates to 1956; a three-year updating and capability enhancement project was completed in 1985. The facility consists of a concrete runway with a pressurized waterjet propulsion system at one end and a cable arrestment system to slow the carriage at the other end; the truss-structure carriage is propelled down the half-mile track on steel rails.

For the heavy rain research program, the facility has been augmented by six trusses spanning the tracks. Suspended from the trusses are 1,590 nozzles that pour simulated heavy rain on the carriage and test model passing underneath.

The 127,000-pound carriage is 70 feet long. The airfoil is mounted on top; the airfoil is a full scale wing section, with working flaps, of a typical modern commercial transport.

A typical test run begins with a burst of high pressure water from the water propulsion system tower, delivered to a waterjet reaction bucket at the rear end of the carriage. The waterjet enters near the top of the bucket and turns through an angle of 170 degrees; the momentum exchange involved in this process provides the thrust to move the carriage.

In two to three seconds, the carriage is accelerated to about 180 miles per hour, simulating takeoff or approach speed. The carriage and test airfoil pass underneath the overhead truss section while hundreds of nozzles bombard the wing model with "rain."

Early results tend to verify the wind tunnel test results for the particular wing being tested—namely that at extremely high rain rates there is a significant loss of lift, a loss that could be critical to flight safety in high-lift conditions such as approach or takeoff.

The current test program is expected to require 60-80 test runs. During the next phase, the project team plans to change the rain rate to determine at what point heavy rain begins to affect performance. The key question is whether rain effect is something to consider only in the most intense downpours or whether it is potentially troublesome even in light or moderate rain.

If extensive track testing shows that heavy rain effect is indeed significant, piloting procedures will have to be revised to take the effect into account. Existing piloting procedures are based on what is known about flying through dry air; no corrections have been

At far left, the heavy rain research carriage is poised for a test run, the test airfoil mounted on top; the tall tower behind the carriage is a waterjet propulsion system, which delivers a burst of high pressure water that creates a reaction that propels the carriage down a half-mile track at simulated approach/takeoff speed. At left below, the carriage is starting to accelerate under waterjet impetus; in the photos below, carriage and airfoil are emerging from the overhead truss section after brief exposure to heavy rain.

developed to account for the effect of heavy rain on airfoil performance. Langley's work will provide an information base for modifying piloting procedures.

The heavy rain research program exemplifies one facet of NASA's broad aeronautical research program: solution of current and predictable aviation problems. Examples include curbing aircraft fuel consumption, reducing airplane and helicopter noise levels, finding ways to alleviate air traffic congestion, and a variety of safety-related investigations, such as anti-icing research, research on fire resistant materials and improved aircraft structures for better passenger protection.

The main thrust of the NASA aeronautical research program is anticipating the longer range needs of future flight and developing applicable technology. Part of this effort involves research of a general nature aimed at advancing aerodynamics, propulsion, materials and structures, aviation electronics and knowledge of the human factors in flight operations. The other part embraces technology development for improving the perform-

Aero-Space Plane

At right is an artist's concept of the National Aero-Space Plane (NASP), a joint Department of Defense/NASA program for development and demonstrations of the technologies essential to a revolutionary class of vehicles that would be capable of taking off or landing horizontally like an airplane, operating in the upper atmosphere at hypersonic speed, or flying directly into Earth orbit.

Successful development of NASP could lead to military or civil single-stage-to-orbit spaceplanes providing rapid access to space with airplane-like flexibility. Because they would not need the extensive facilities and lengthy countdown operations needed for vertical space launches, such craft promise dramatically lower costs of delivering payloads to orbit.

The technologies are also applicable to hypersonic (above 4,000 miles per hour) military interceptors. The technologies could eventually be incorporated in future generations of hypersonic military transports, but current research focus for advanced transports is on the supersonic Mach 2-3 (1,300 to 2,000 miles per hour) flight regime rather than hypersonic flight.

The current phase of the NASP effort, which began in 1986, involves development of key technologies in propulsion, aerodynamics, advanced structures , high temperature resistant materials and computational fluid dynamics, or CFD (computer simulation). Since there are no wind tunnels capable of testing propulsion models above Mach 8, it is necessary to employ CFD techniques to determine how hypersonic airflows will affect the NASP vehicle.

A planned technology demonstration phase involves construction and flight test of an experimental vehicle designated X-30. In mid-summer 1989, the pace of ground development and flight test schedules were under review.

The NASP program is a nationwide effort involving government researchers at the USAF's Aeronautical Systems Division, Air Force Weapons Laboratory and Arnold Engineering and Development Center; NASA's Ames, Langley and Lewis Research Centers; the Naval Surface Weapons Center; and the Department of Energy Los Alamos Laboratory.

A team of private industry contractors is sharing development costs with the government and operating as a non-competitive consortium to share research data, keep costs down and speed technology development. Airframe development contractors include General Dynamics Corporation, McDonnell Douglas Corporation, and North American Aircraft Operations Division of Rockwell International. Engine contractors are Pratt & Whitney Division of United Technologies and Rocketdyne Division of Rockwell International. Aerojet Techsystems and Marquardt Company are developing and building hypersonic engine test facilities for the NASP program.

Airflow Research

Shown at right is the NASA F-16XL research aircraft being flown at Ames-Dryden Flight Research Facility to evaluate concepts designed to improve wing airflow in supersonic flight.

Current aircraft wings have turbulent airflow over a large portion of their upper surfaces. Friction between the turbulent air and the airplane skin causes aerodynamic drag that reduces aircraft fuel efficiency.

If the airflow could be maintained throughout a flight in "laminar" condition (smooth, or non-turbulent) exceptional gains in fuel efficiency could be realized. NASA has established the feasibility of reducing fuel consumption by laminar flow control technology that employs suction to keep the airflow smooth. Flight tests of this concept have been limited to subsonic speeds.

Initial experiments with the F-16XL involve test of the suction mode of laminar flow control at supersonic speeds up to Mach 2. Instruments acquire in-flight data from an experimental wing section perforated with millions of tiny, laser-cut holes connected to an air pump in the fuselage; the pump's suction smooths the turbulent air, establishing laminar flow. Information from F-16XL testing is expected to aid in the design of future high speed aircraft, including civil supersonic transports whose fuel consumption is critical to their economic viability.

NASA is using two General Dynamics-built F-16XLs on loan from the Air Force. The aircraft feature a NASA "cranked arrow" design, so called because of the arrowhead-like shape of the wing, which sweeps sharply rearward behind the cockpit, then flares out to a more moderate sweep angle. This design was proposed by NASA for the U.S. supersonic transport program of the 1960s. NASA later provided design support to General Dynamics in the development of a company-funded prototype F-16XL.

High-angle Research

At right is a specially equipped and instrumented NASA F-18 research aircraft being flown at Ames-Dryden Flight Research Facility to investigate the behavior of high-performance aircraft flying at high angles of attack.

An airplane's angle of attack is the angle between the wing and the air through which it moves. At high angles of attack, airflow around the airplane becomes extremely complex and accurate information about such airflows is scant.

NASA is seeking new understanding of high-angle airflows through flight research supported by wind tunnel data and computer simulations. Expanded knowledge of aircraft aerodynamics at angles of attack above 45 degrees could enable accurate prediction of the complex airflow interactions. These predictions could provide design criteria to increase significantly the maneuverability of high performance aircraft and prevent accidents related to high angles of attack.

The basic aim of the High Angle of Attack Research Program is to create a data base and develop methods that will permit more efficient design of high performance aircraft, thereby minimizing costly post-production fixes. Managed by Ames-Dryden Flight Research Facility, the program is a cooperative effort of Ames, Langley and Lewis Research Centers. Ames is developing computer simulation data, Ames and Langley are both providing wind tunnel data, and Lewis is handling propulsion-related research.

The F-18 has been conducting flight evaluations since April 1987 and by mid-summer 1989 it had made approximately 90 flights at angles of attack up to 55 degrees. During that span, research emphasis was on flow visualization studies. The smoke trails streaming from the airplane's nose in the upper right photo represent one flow visualization technique; these trails, created by a nose-mounted smoke generator, visualize vortex flows, mini-tornados swirling around parts of the aircraft that increase lift and possibly can be utilized for aircraft control at high angles of attack, where most aircraft handle poorly. Location and apparent intensity of the vortices is recorded by five cameras in the F-18's wingtips, tailtips and fuselage.

Another technique involves release of a dyed liquid from the nose section to "paint" on the airplane's skin a pattern of streamlines representing airflows around the wing and fuselage.

In the fall of 1989, F-18 flights will be suspended while the airplane undergoes modification for a new phase of the program: testing control and maneuverability through thrust vectoring, or deflecting engine exhaust. McDonnell Douglas Corporation, original builder of

the F-18, is developing a thrust vectoring control system for installation in the NASA research aircraft and flight tests beginning in the spring of 1990.

A parallel high angle of attack research project under way at Ames-Dryden is a joint Air Force/NASA X-29 Advanced Technology Demonstration program. The X-29 program is intended to demonstrate a variety of advanced technologies that collectively can make possible construction of smaller, lighter and more efficient aircraft without sacrificing performance. Among the technologies are a unique forward-swept wing made of composite materials; rotating canards that replace conventional horizontal tail surfaces to control pitch; and an advanced digital flight control system that stabilizes the aircraft by adjusting the wing trailing edges, canards and other control surfaces more than 40 times a second. The USAF Aeronautical Systems Division manages the X-29 program and Ames-Dryden is responsible for flight test activity.

Grumman Aerospace Corporation built two X-29s, the first of which completed a 242-flight test program in December 1988 to evaluate the forward-swept wing and other technologies. The Number Two X-29 (right), identical in configuration to Number One but equipped with special instrumentation and a spin-recovery parachute, is intended to investigate the maneuvering capability of the forward-swept wing/canard design at extremely high angles of attack, up to 70 degrees (the Number One X-29 was never flown beyond a 22 degree angle). The Number Two craft made its initial flight on May 23, 1989; the flight test program contemplates 70 flights through September 1990.

Vortex Flap Research

The airplane in both photos is a Langley Research Center F-106B equipped with a research innovation known as the vortex flap, visible in the photos as an extension of the wing leading edge pointed sharply downward.

The vortex flap is a drag reduction device intended to increase the survivability of military fighters by increasing their maneuverability in air combat action. Since drag reduction also reduces fuel consumption, the technology is applicable to any highly-swept-wing airplane, including civil transport aircraft.

The vortex is a tornado-like swirl of air that reduces drag by creating a thrust component. Computer predictions suggest that drag reductions of as much a 20 percent can be attained by shifting a wing's vortex system to the forward edge of the wing. This is accomplished by mounting vortex flaps along the full length of the wing leading edge and deflecting them downward; the flaps of the research craft are ground-adjustable to 20, 30, 40 and 50 degrees.

To allow visualization of the vortex field, researchers have equipped the black-painted wing of the F-106B with "flow cones," tufts that indicate the motion of air over the vortex flap and the normal wing surfaces. The flow is observed by means of video cameras mounted on the upper fuselage and inlet cowl.

At midyear 1989 the F-106B had flown about 20 flights at speeds up to Mach 1 and altitudes to 40,000 feet; on all flights the flaps were deflected to 40 degrees. The research program contemplates flights over the full transonic maneuvering envelope of the airplane to gather flap-aided maneuvering performance data with flap deflections at other angles.

Takeoff Monitor

A modern jetliner's cockpit has informational displays showing just about everything a pilot needs to know, but the decision to lift the plane off the runway is still a "seat of the pants" judgment on the part of the pilot. There is no simple "Go/no go" advisory during a takeoff roll.

To improve safety in the decision-making process, Langley Research Center is developing a Takeoff Performance Monitoring System (TOPMS) that wraps up in a visual package all the data necessary for a successful takeoff.

TOPMS is a computer program coupled to a display system that shows where on the runway important takeoff events will occur and whether the takeoff run should be continued or aborted. The photo shows a TOPMS display in a Langley transport systems research simulator. TOPMS displays a runway scene with coded symbols relating such information as airplane speed at any time during the takeoff run, whether acceleration is satisfactory or not, engine status, the predicted point on the runway where liftoff will take place, and the point where the airplane can be stopped with maximum braking.

TOPMS goes one big step further and summarizes all this data in a "Stop" or "Go" indicator.

TOPMS has been extensively tested in the simulator by scores of pilots, civil and military, who have without exception declared it would be very useful in operational service. The system has also been checked out in more than 50 real takeoffs with Langley's Transport Systems Research Vehicle, a Boeing 737 jetliner used in developmental tests of advanced airborne systems and flight procedures. TOPMS flight tests included normal takeoffs, high altitude takeoffs and high-speed aborts; the system performed satisfactorily in all situations.

Commercial Use of Space

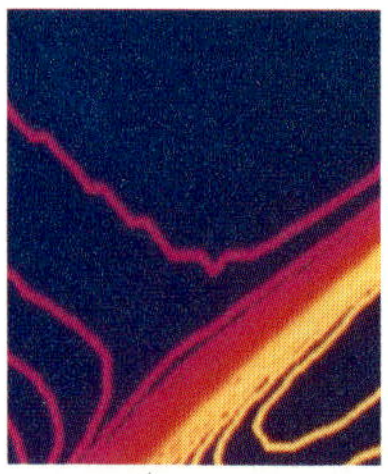

On Space Shuttle mission STS-29, launched in March 1989, the Orbiter Discovery carried a package of protein crystal growth experiments aimed toward eventual production of powerful new drugs to fight cancer and other diseases. Also in March, another microgravity research package was sent aloft on the first U.S.-licensed commercial sounding rocket flight. And on the STS-30 Shuttle mission launched in May, a small microgravity laboratory flew in the middeck of Orbiter Atlantis.

These are examples of the accelerating pace of commercial space activity. There are many more.

NASA's effort to stimulate interest and investment in space-related ventures is gaining momentum

By mid-1989, more than half of the 50 largest U.S. industrial corporations were participating in one or more of NASA's programs for stimulating commercial space research. Some are interested in the potential of orbital space as a unique place to conduct research and develop new or improved processes for manufacturing commercial products. Some are pioneering private development of commercially marketable space research facilities and equipment. Others are pursuing commercial opportunities in launch and orbital services.

Some of these companies are old-line aerospace firms—Boeing, McDonnell Douglas, Rockwell International, United Technologies. But the broadening interest in the commercial space potential is underlined by the fact that most of these firms are non-aerospace companies, familiar names most usually associated with consumer products or services—3M, Amoco, du Pont, and Eastman Kodak, to name a few.

There are several different types of cooperative arrangements wherein NASA and industry work together to further U.S. leadership in commercial development of space for the national good. The two principal arrangements are the Joint Endeavor Agreement (JEA) and the Centers for the Commercial Development of Space.

The JEA is a program through which U.S. companies commit private resources to build and conduct space experiments and NASA sponsors Space Shuttle flight opportunities for them. In mid-1989, NASA had five such agreements with requirements for more than 70 flights. In the months since the resumption of Shuttle service with the flight of STS-26 in September 1988, the Office of Commercial Programs has identified eight potential new JEA partners.

The Centers for the Commercial Development of Space (CCDS) are NASA-sponsored, university-based not-for-profit research consortia composed of industrial firms, academic institutions and government organizations. Each of the centers focuses on a particular field of space endeavor that offers potential for commercial development and thus attracts industry interest and participation. Since the program was established in 1985, the number of centers has grown to 16 and the number of participating industrial firms to 128. Six of the centers are focusing on materials processing research; two are specializing in commercial remote sensing operations, two in space power, two in bioscience, two in robotics and one each in space propulsion and materials for space structures. The Center for Macromolecular Crystallography at the University of Alabama-Birmingham became the first CCDS to conduct an orbital flight experiment in 1988, aboard Shuttle flight STS-26.

A concept for mining helium on the moon by robotic vehicles exemplifies the type of innovative work being done by the NASA-sponsored Centers for the Commercial Development of Space (CCDS), which are research teams that include industry, academic and government participants. The lunar concept was developed by the Wisconsin Center for Space Automation and Robotics.

Among other areas of progress,

• NASA and SPACEHAB, Inc. signed a Space Systems Development Agreement for orbital flight in the Shuttle of the company's commercially-developed and manufactured pressurized module, designed for research requiring man-tended access to space.

• NASA signed three new facilities use agreements with manufacturers of expendable launch vehicles.

• NASA announced a program of funding support for commercial sounding rocket flights involving investigation of industrial space applications developed by the CCDS.

• NASA signed the first agreement for commercial use of Space Shuttle External Tanks; three additional proposals were being further evaluated.

• A Commercial Programs Advisory Committee that includes 18 industry chief executives and university presidents was established to help determine a national course for commercial development of space.

• A study of issues and potential commercial space objectives, conducted under NASA contract by the American Institute of Aeronautics and Astronautics, was completed.

• An Operations Management Council, intended to improve the effectiveness of CCDS operation, was established.

• The Office of Commercial Programs developed a flight plan to maximize opportunities for CCDS payloads and JEA experiments.

• A Commercial Space Infrastructure Policy, Criteria and Procedures document for evaluating industry proposals was drawn up.

These developments are amplified in the following pages.

Microgravity Research

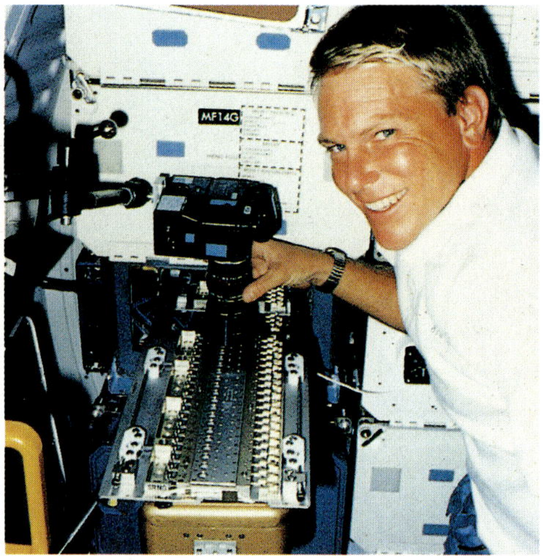

After a 30-month hiatus, orbital research on materials processing in microgravity resumed on Space Shuttle flight STS-26 in September 1988. On board the Shuttle Orbiter Discovery were two experiments, one an investigation seeking confirmation that the space environment will allow production of protein crystals of higher quality than can be made on Earth, the other part of a continuing study of how organic thin films behave in microgravity.

Protein crystal growth is among the most promising of the near term applications of microgravity processing. Space grown proteins may become vitally important research tools for scientists working to develop powerful new drugs to combat cancer, high blood pressure, organ transplant rejection, rheumatoid arthritis and many other disorders.

At upper right, astronaut George Nelson is monitoring protein growth experiments aboard STS-26. In a NASA-built apparatus carried in the Orbiter's middeck area, tiny crystals form in liquid droplets within 60 experiment chambers. Postflight photos, such as the one at right, document the quality of the crystals. The STS-26 protein growth experiment was sponsored by the Center for Macromolecular Crystallography (CMC) at the University of Alabama-Birmingham, one of 16 NASA Centers for the Commercial Development of Space. The team of investigators included scientists from CMC, Upjohn Company, E.I. du Pont de Nemours, the Merck Institute for Therapeutic Research and the Schering Corporation.

The other microgravity experiment aboard STS-26 was sponsored by 3M Company's Space Research and Applications Laboratory as part of a long range program of basic research expected to lead to creation of new technologies and new products. Called PVTOS, for Physical Vapor Transport of Organic Solids, the experiment was aimed at producing organic thin films with ordered crystalline structures for study of their optical, electrical and chemical properties. The apparatus consisted of nine independent cells, each containing a sample of organic material that was vaporized during orbital flight to form a thin film. The films were returned to Earth for studies that could eventually lead to on-Earth or on-orbit production of specialized commercial thin films. At right, PVTOS principal investigator Dr. Mark Debe is using an infrared spectrometer, the device used for postflight characterization of the films produced on STS-26.

On flight STS-29, flown in March, 1989, CMC continued its protein crystal growth research, this time growing crystals from 60 different material samples. The team of investigators included scientists from eight industrial affiliates of CMC: du Pont, Upjohn, Schering, Merck Institute, Eastman Kodak, Eli Lilly & Company, Smith, Kline and French, and Biocryst Limited. In postflight analysis, the space grown crystals are being compared for shape, quality and composition with Earth grown crystals.

Flight STS-30, May 1989, employed the Rockwell International Fluids Experiment Apparatus (FEA), a small microgravity laboratory (right) about the size of a 19-inch TV designed to handle a range of processing applications, including liquid chemistry, fluid physics, thermodynamics, biological cell culturing and crystal growth. The STS-30 FEA experiment was one known as the Float Zone Refining Experiment, jointly developed by Rockwell's Science Center and Indium Corporation of America, a leading refiner of the metal indium for a broad variety of applications.

The research involved on-orbit refining of indium by a "float zone" process in which impurities are progressively removed. Indium Corporation hopes to determine through orbital research the coefficients of separation of elements, which cannot be determined by Earth-based research; such knowledge could lead to development of Earth-based refining processes for producing ultrahigh purity metals. Rockwell Science Center and Indium are jointly conducting postflight analysis.

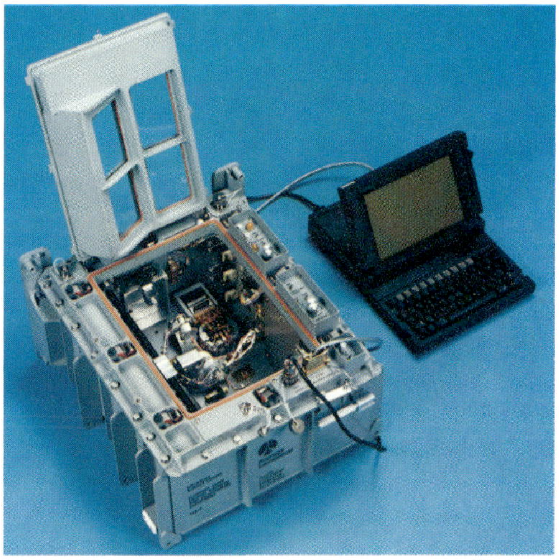

Launch Services

At right is the Starfire sounding rocket, built by Space Services, Inc., Houston, Texas, lifting off from the White Sands Missile Range, New Mexico on March 29, 1989. The event marked the first flight of a U.S.-licensed commercial sounding rocket.

The Starfire flew a 15-minute suborbital flight path carrying six materials science experiments that were exposed to microgravity conditions for about eight minutes. The mission was designated Consort 1 for its sponsor, the Consortium for Materials Development in Space. Both the sounding rocket and its payload operated flawlessly; at lower right, members of the mission team examine a segment of the payload.

The success of Consort 1, a pilot project in which NASA financed the purchase of commercial launch services and payload integration, prompted NASA to announce support of a series of commercial sounding rocket flights that will provide opportunities—in addition to experiment packages on Shuttle flights—for investigations of industrial space applications by the 16 NASA-sponsored Centers for Commercial Development of Space. The first launch under the new program is scheduled for late 1989.

In line with the government policy of encouraging and assisting the growth of the U.S. commercial launch industry, NASA signed four new agreements in 1988/1989:

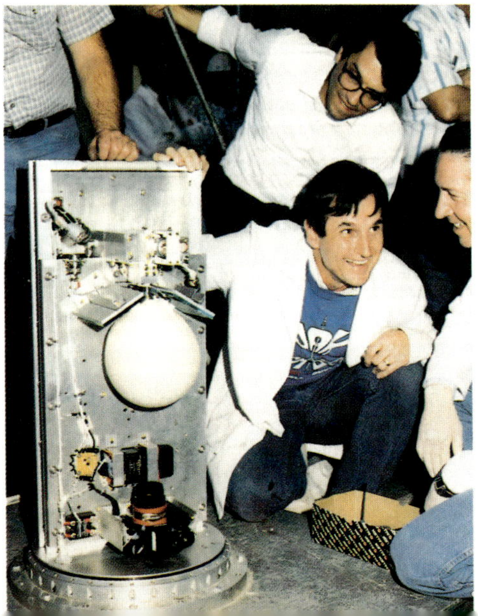

• In September 1988, NASA and McDonnell Douglas Astronautics Company concluded an arrangement providing for the firm's use of facilities at Kennedy Space Center (KSC) and technical assistance by Goddard Space Flight Center in support of commercial launches of the McDonnell Douglas Delta vehicles (right center). In July 1989, NASA selected McDonnell Douglas for negotiations leading to a contract for medium class launch vehicle

services for three firm missions in 1992/93 and 12 optional missions.

• In October 1988, a similar agreement was signed with Martin Marietta Commercial Titan, Inc. granting the company access to NASA-managed facilities at KSC for launches of the Commercial Titan (above right), a version of the Air Force-developed Titan III.

• In December 1988, NASA granted LTV Missiles and Electronics Group exclusive rights to produce and launch the NASA-developed Scout (right) on a commercial basis. The agreement gives LTV use of NASA-controlled production tooling and special test equipment needed for Scout manufacture; additionally, it allows use of Scout launch support facilities at NASA Wallops Flight Facility and at Vandenberg Air Force Base.

These agreements followed the original 1987 NASA commercial launch service support agreement with General Dynamics Corporation, whereby operation of the government-developed Atlas Centaur was transferred to the company and company access to KSC launch facilities was granted. In 1988, NASA awarded General Dynamics a contract for Atlas Centaur launch services for a new family of environmental satellites, the first government procurement of commercial launch services.

Commercial Space Developments

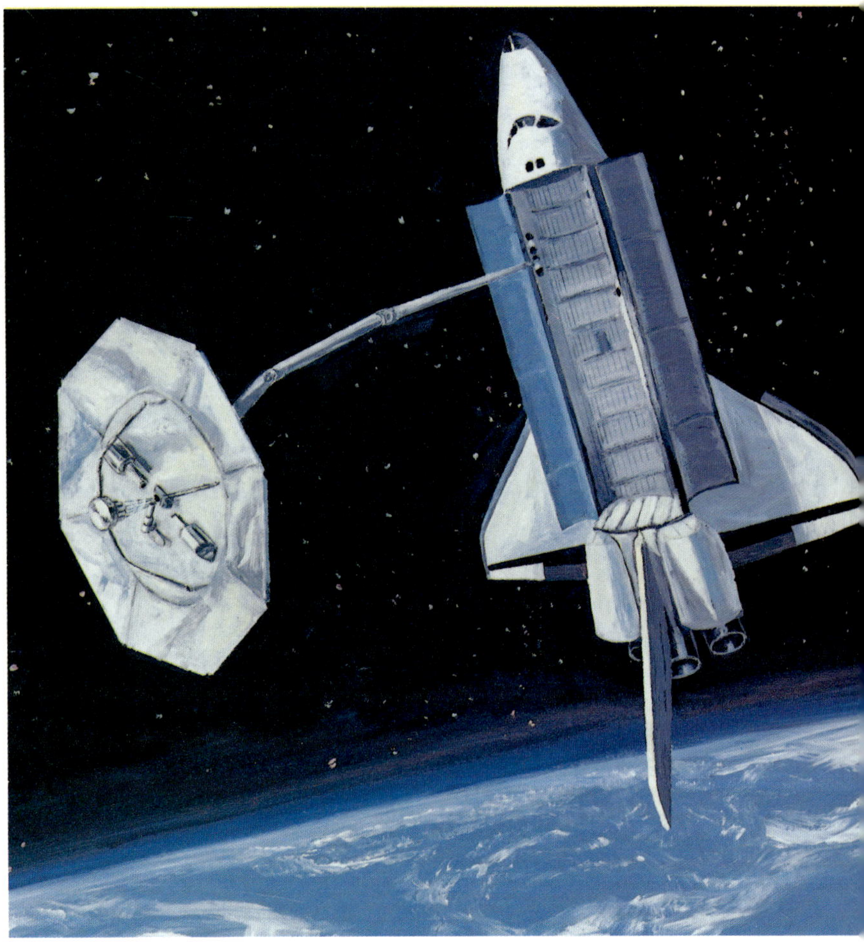

Epitaxy is a space lexicon term that will be heard more and more frequently; it is the growth of crystals in a special atom-by-atom, layer-by-layer manner to produce varying crystalline structures.

Epitaxial growth under high vacuum conditions can produce crystalline thin films of higher quality and purity than can be grown on Earth. Such films offer promise of important technological advances in superspeed computers, lasers, communications and infrared devices, and many other high technology microelectronic applications.

NASA considers epitaxial thin film growth so important that one of the agency's 16 Centers for the Commercial Development of Space (CCDS) is devoted specifically to this area of research: the Space Vacuum Epitaxy Center (SVEC) at the University of Houston.

Orbital epitaxial research will be made possible by a Wake Shield Facility, a 10-foot-diameter disc carried in the Space Shuttle Orbiter. The Facility will be extracted from the Orbiter's payload bay by the remote manipulator arm and held aloft (right) to sweep an orbital wake, creating an ultravacuum region behind the disc. Epitaxial thin films will be grown on the disc in ultravacuum conditions, then the Facility will be retracted and stowed in the payload bay. Four such flights are planned, beginning in 1991.

SVEC has awarded a contract to Space Industries, Inc. for a major portion of the design and construction of the Wake Shield Facility. Initial experiments are being readied by SVEC and its affiliates—AT&T, Electro-Optek Corporation, Perkin-Elmer Corporation, Rockwell International, Instruments S.A., Inc., the University of Illinois and the U.S. Army.

In other areas of commercial space activity, NASA signed agreements involving development and flight of two Shuttle-related systems.

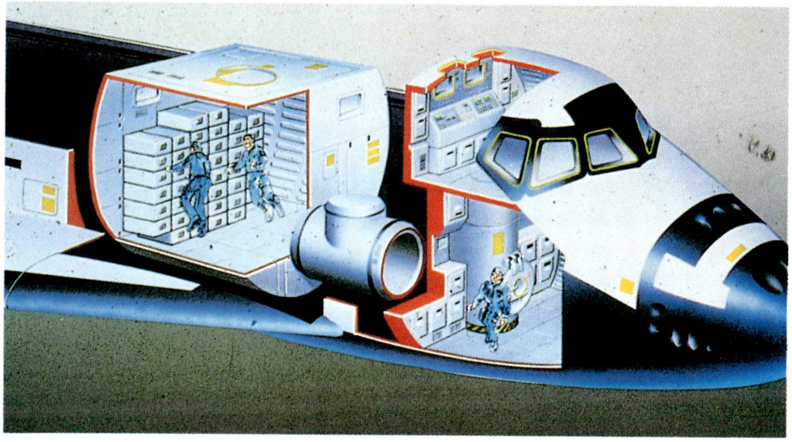

One agreement, with SPACEHAB, Inc., provides for six Shuttle flights, beginning in 1991, of the company's privately developed pressurized module designed to augment the existing pressurized volume of the Shuttle Orbiter's middeck; the SPACEHAB module is shown in the Orbiter's payload bay at upper right and in a cutaway view at lower right. The company is marketing access to the 10 by 13-foot module on a commercial basis. SPACEHAB has selected McDonnell Douglas Astronautics Company as prime contractor for final design and construction of the modules; subcontractors include United Technologies' Hamilton Standard Division and Italy's Aeritalia.

The other agreement, with the University Corporation for Atmospheric Research (UCAR), is for use of Space Shuttle External Tanks (left below) as laboratories. UCAR intends to conduct scientific research through experiments mounted in a large segment of the mammoth tank known as the intertank area, a 5,000-cubic-foot volume between the fuel and oxidizer tanks. These experiments will be conducted during the suborbital trajectory of the tank, after it burns off its propellants and is jettisoned from the Shuttle. Under the agreement, NASA will give UCAR intertank space on up to five External Tanks.

Another development represents the first spinoff from CCDS research. The Center for Commercial Development of Space Power at Auburn University and Maxwell Laboratories, San Diego, California jointly developed a space power supply device that also has wide Earth-use applicability.

For either space or Earth applications, the system is lighter and more compact, more reliable and more efficient than existing systems that do the same job. It is expected to be used on future space power devices, but the commercial spinoff is already in production. The commercial product, about the size of a stereo receiver, transforms and conditions large voltages to charge capacitors used in such medical/industrial equipment as lasers, x-ray systems, radars and microwave communications systems. Maxwell Laboratories will produce and market the product and share revenues with the Auburn CCDS.

Still another NASA/industry agreement—this one with Corabi International Telemetrics—involves a "reverse spinoff" whereby Corabi will adapt for Space Station use its terrestrial technology that enables medical specialists to view remotely and analyze biological and other materials via high resolution video images. Corabi is interested in providing—on a commercial basis—a Space Station telerobotic workstation that transmits a body or specimen image to a ground station anywhere in the world for patient analysis. The agreement is expected to lead to a Corabi proposal for proof-of-concept Shuttle-based tests of a prototype system.

Space Operations: In Earth Orbit and Beyond

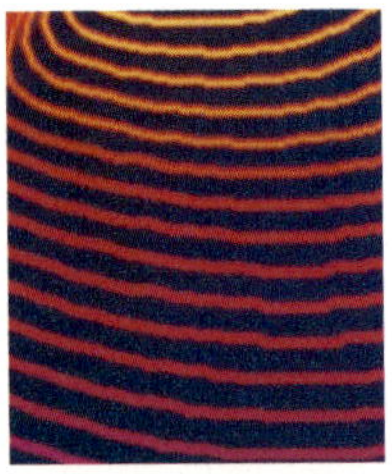

"Our goal is nothing less than to establish the United states as the preeminent spacefaring nation," President Bush declared at a July 20, 1989 ceremony commemorating the 20th anniversary of the initial Apollo manned lunar landing.

The President keyed attainment of the goal to three space giant steps: completion of the Space Station *Freedom*, which he termed the "critical next step to all our space endeavors;" next, a manned lunar outpost; then, at an untargeted time in the 21st century, "a mission into tomorrow, a journey to another planet, a manned mission to Mars."

President Bush directed the National Space Council to review options and costs for the lunar/Mars missions and report recommendations.

Space Station Freedom and other planned technology advancements exemplify NASA steps toward expanding human presence in space

Accomplishment of the lunar outpost/Mars expedition goals demands a broader science and technology base than currently exists. NASA is working toward that end with its Pathfinder program, a technology initiative that will allow development of critical capabilities to enable a wide range of manned and unmanned missions beyond Earth orbit.

Pathfinder complements the ongoing Civil Space Technology Initiative (CSTI). Where CSTI aims at developing technologies related to space operations in low Earth orbit, Pathfinder's focus is on emerging technologies that would make possible advanced solar system exploration with manned and robotic spacecraft.

The Pathfinder research and development effort is divided into four major program areas:

- *Surface Exploration*, involving development of critical technologies for exploring planetary surface, with principal focus on the beginning of extended human operations on the Moon and the exploration of Mars.
- *In-Space Operations*, which embraces a wide assortment of key technology advances that will significantly expand the U.S. capability for space operations in support of solar system exploration.
- *Space Transfer*, development of capabilities to allow movement from one orbit to another or from one solar system body to another.

• *Humans in Space,* involving R&D on essential technologies to permit humans to live and work for extended periods in solar system environments beyond Earth.

In addition, Pathfinder embraces detailed study of extraterrestrial missions, such as human-crewed missions to the Moon or Mars, robotic exploration of the solar system, or robotic "precursors" to human missions, for example, a mission involving robotic Mars-roving and return of a Martian sample to Earth, an essential preliminary to a manned mission to the planet. *(Continued)*

In the Pathfinder program, NASA is developing critical technologies that will enable such 21st century missions as a lunar outpost (above), a manned mission to Mars (left), and advanced robotic exploration of the solar system.

Space Operations: In Earth Orbit and Beyond

(Continued)

An example of a type of vehicle that would be needed to make possible the giant step missions contemplated for the 21st century is the autonomous planetary lander. Autonomous in this sense means without help from Earth; the time it takes a communications signal to make the round trip between Earth and a planet, even at the speed of light, rules out real-time Earth-based control of planetary landings. The vehicle must be capable of landing safely, avoiding such surface hazards as rocks and slopes, and accurately, close to preplanned targets of prime scientific interest, with intelligence provided by onboard systems.

Development of technology for an autonomous lander is one of the subelements of the Pathfinder Surface Exploration program. Other subelements in that area of effort include a planetary rover for moving about on a solar system surface, either robotically or under human control; technology for acquiring, analyzing, storing and returning to Earth surface and subsurface samples; power systems capable of supporting manned operations on the Moon or Mars; and photonics technology for extremely powerful information processing systems, and for computers and networks with extremely high degrees of fault tolerance.

Concept of an autonomous Mars lander, meaning one capable of avoiding surface hazards and landing at a precise spot without the help of Earth-based control or advisories.

The In-Space Operations program area includes development of technologies for rendezvous and docking in lunar/planetary orbits; in-space assembly and construction of large, massive facilities and complex spacecraft; an orbital fueling station; finding ways to use nonterrestrial resources to reduce Earth-support requirements; and development of optical communications systems for the extremely high rates of data transmissions that will be required.

Another program area is Space Transfer, which involves technologies for moving from one point in space or one orbit to another, including Earth-Moon transportation and transfers among the planets and other bodies of the solar system.

This part of the Pathfinder program is subdivided into three main areas of effort: technology for chemical propulsion of transfer vehicles and planetary landers; nuclear electric propulsion systems for propulsion of cargo vehicles, piloted or robotic; and "aerobraking," a technique in which the atmospheres of Earth or other solar system bodies are employed to decelerate a spacecraft in order to attain the requisite velocity for orbiting the body.

The remaining Pathfinder program area is Humans in Space, which has several subelements. One is directed toward providing a technology base for advanced extravehicular space suits that will allow humans to operate for long periods outside the protection of

their pressurized spacecraft or surface habitats. Another, called Space Human Factors, seeks a significant increase in understanding of human performance and behavior in long durations space missions. A related effort, Human Performance, involves investigations of human tolerances to such space phenomena as weightlessness, artificial gravity and radiation. Other subelements include crew protective systems and regenerative life support systems, where water, air and other resources are reused to reduce the mass of consumables—hence the cost of resupply—required for extended duration human space activities.

An orbiting transportation depot for transfer vehicles exemplifies the type of large structure being contemplated for in-space assembly and construction.

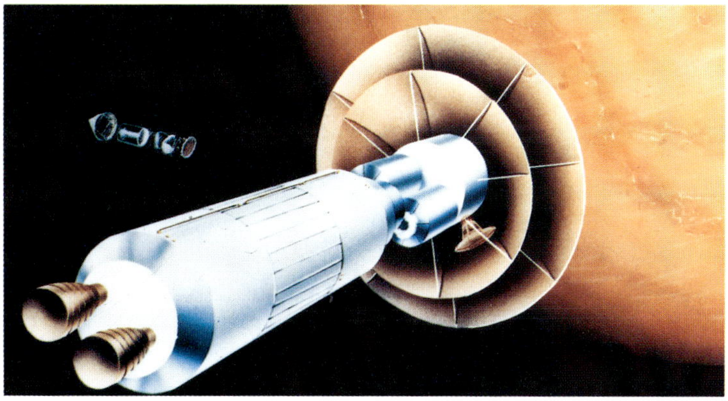

A major Pathfinder aim is development of technologies for transfer vehicles, such as this one approaching Mars.

Shuttle Operations

At right, the first stage of the Inertial Upper Stage (IUS) booster is separating from the Tracking and Data Satellite (TDRS) as the satellite nears its planned position in geosynchronous orbit. Below, the booster's second stage has dropped off and the satellite is fully "unfolded" with its solar panels and antennas deployed.

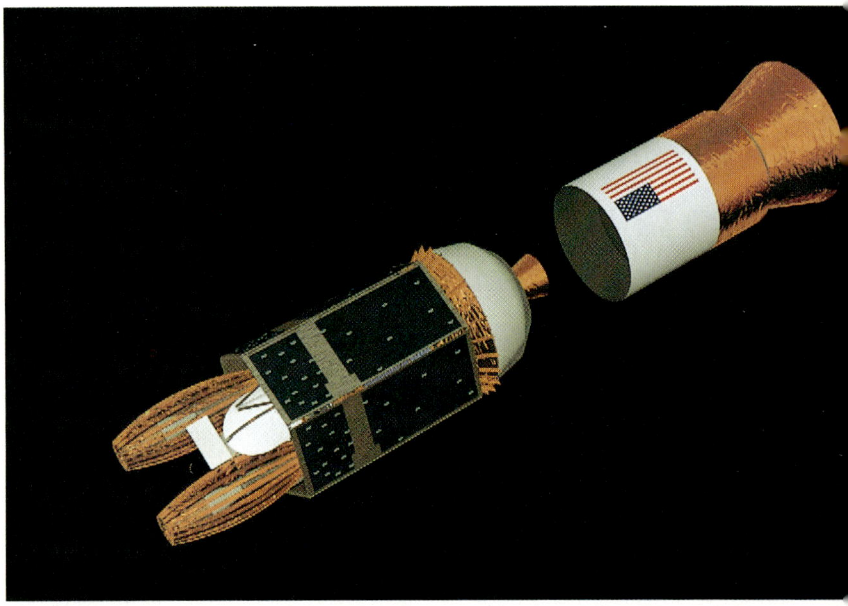

Delivery of TDRS-D was the primary assignment of Shuttle flight STS-29, launched March 13. It was the first of six Shuttle flights planned for 1989, to be followed by nine in 1990 and eight in 1991. In 1992, when the new Shuttle Orbiter *Endeavour* joins the fleet and NASA once again has four Orbiters, flight frequency will increase to 12, then build up to 14 flights in 1995 when assembly of the Space Station *Freedom* begins.

STS-29 was flown by the Orbiter *Discovery*, which was making its eighth flight overall and its second of the post-Challenger "new beginning" era. Delivery of TDRS-D completed the three-satellite constellation known as the Tracking and Data Relay Satellite System (TDRSS), which will increase communications between Earth-orbiting spacecraft and ground stations by 15-85 percent per orbit, making possible a much higher rate of data flow. CONTEL (Continental Telecom Inc.) owns and operates the TDRSS for NASA; the satellites are built by the Defense and Space Systems Group of TRW Inc.

Other activity during the five-day STS-29 mission involved experiments with a number of secondary payloads. An experiment known as SHARE, for Space Station Heat Pipe Advanced Radiator Element, tested a possible cooling system for Space Station *Freedom*. A Chromex experiment sought to determine whether the roots of a plant in microgravity would develop similarly to Earth roots. Sixty different protein crystal growth experiments were conducted (see page 35) and there were two student experiments on the effects of weightlessness. Another secondary payload was an IMAX camera, used in a joint NASA/National Air and Space Museum project to document space activities with specially designed 70 millimeter cameras; at right above, "upside down" mission pilot John E. Blaha is using the overhead window on *Discovery's* aft deck to capture IMAX Earth scenes.

The primary objective of STS-30, Orbiter *Atlantis,* launched May 4, was to send the Magellan radar imaging spacecraft on its way to Venus, an assignment successfully carried out (see page 8); at right, Magellan—affixed to its IUS booster—has just departed the Orbiter.

Among secondary objectives of STS-29 was one of a series of AMOS tests. AMOS is the Air Force Maui (Hawaii) Optical Site, a tracking station. Tests allow AMOS ground-based electro-optical sensors to collect Orbiter imagery and signature data during cooperative overflights; the data provides scientific information and additionally serves to support calibration of AMOS sensors.

Another secondary objective was a Mesoscale Lightning Experiment, designed to obtain nighttime images—over an extremely large geographical area—of lightning; the tests are intended to develop understanding of what effects lightning discharges have on each other, on storms, microbursts and wind patterns, and on other interrelationships. A third objective involved microgravity research by an industry team gathering data, with astronaut assistance, from a microgravity laboratory known as the Fluids Experiment Apparatus (see page 35). At right, mission specialist Mary L. Cleave aims a data recording camcorder into the apparatus during a materials science test.

Orbiter *Columbia* made its first post-Challenger flight on August 8, carrying a classified Department of Defense payload.

Advanced Launch Vehicles

At upper right is an artist's conception of Shuttle-C, an expendable cargo delivery evolution of the Space Shuttle capable of lifting 50 to 75-ton payloads to low Earth orbit.

Shuttle-C is intended as a heavy lift enhancement of the Space Transportation System available at relatively low cost because the design philosophy takes advantage of existing components and facilities to curb costs. The vehicle employs without change the Space Shuttle's two Solid Rocket Boosters and the External Tank, and it would use the same ground facilities as the Space Shuttle.

The winged Orbiter, not required because the system is unmanned and does not return to Earth, is replaced by the Shuttle-C Cargo Element (SCE), which consists of an 82-foot-long cylindrical payload carrier and an aft fuselage housing the liquid-fueled main engines, which are previously flown Space Shuttle main engines. Depending on payload requirements, Shuttle-C would have either two or three main engines.

Payload capabilities—for Shuttle-C flights to Space Station *Freedom*—range from 100,000 pounds with two main engines to 170,000 pounds with three main engines plus a booster powered by the Advanced Solid Rocket Motor. This compares with the Space Shuttle's capability of delivering 39,500 pounds to the Space Station.

Shuttle-C is in study status, not yet approved for development; studies are being conducted by United Technologies' Space Flight Systems Division, Martin Marietta Aerospace and Rockwell International, prime contractor for the Space Shuttle Orbiter. Because of the high commonality with existing flight components and ground facilities, Shuttle-C can be developed economically and in a relatively short time, available for a first launch with four years of official authorization.

NASA is also engaged, jointly with the Department of Defense, in development of the Advanced Launch System (ALS), described as a "flexible, robust, reliable and responsive launch system capable of significantly lower costs of getting payloads to space." A primary goal is to reduce the cost of delivering a pound of payload to orbit from its current level of about $3,000 to approximately one-tenth that. Additional goals are systems reliability and resiliency, reduced system complexity and streamlined operations procedures.

The Department of Defense is managing systems engineering and integration, the basic vehicle, the payload module and logistics. NASA's role is to develop advanced liquid engine technologies for the ALS and provide vehicle technologies in NASA's areas of expertise: aerodynamics, materials, structures and operations. Marshall Space Flight Center (MSFC) has the lead responsibility for NASA's ALS activities.

As currently envisioned, ALS would consist of a core stage powered by liquid hydrogen/liquid oxygen-fueled engines, with either a solid rocket or a liquid-fueled rocket booster stage, plus a payload compartment or canister. Both core and booster may be expendable or partially recoverable, and another option is a fully reusable glideback/flyback booster.

One facet of the ALS program involves testing of an Advanced Recovery System (ARS) being developed by Pioneer Aerospace under MSFC management. Shown at lower right in Ames Research Center's 80 x 120 foot wind tunnel, the free world's largest, the ARS is being developed to demonstrate the applicability of high glide deployable wing technology to the soft-landing recovery of large, valuable space assets, in particular the propulsion and avionics modules employed in advanced launch systems. Operating costs of any launch system can be significantly reduced by recovering,

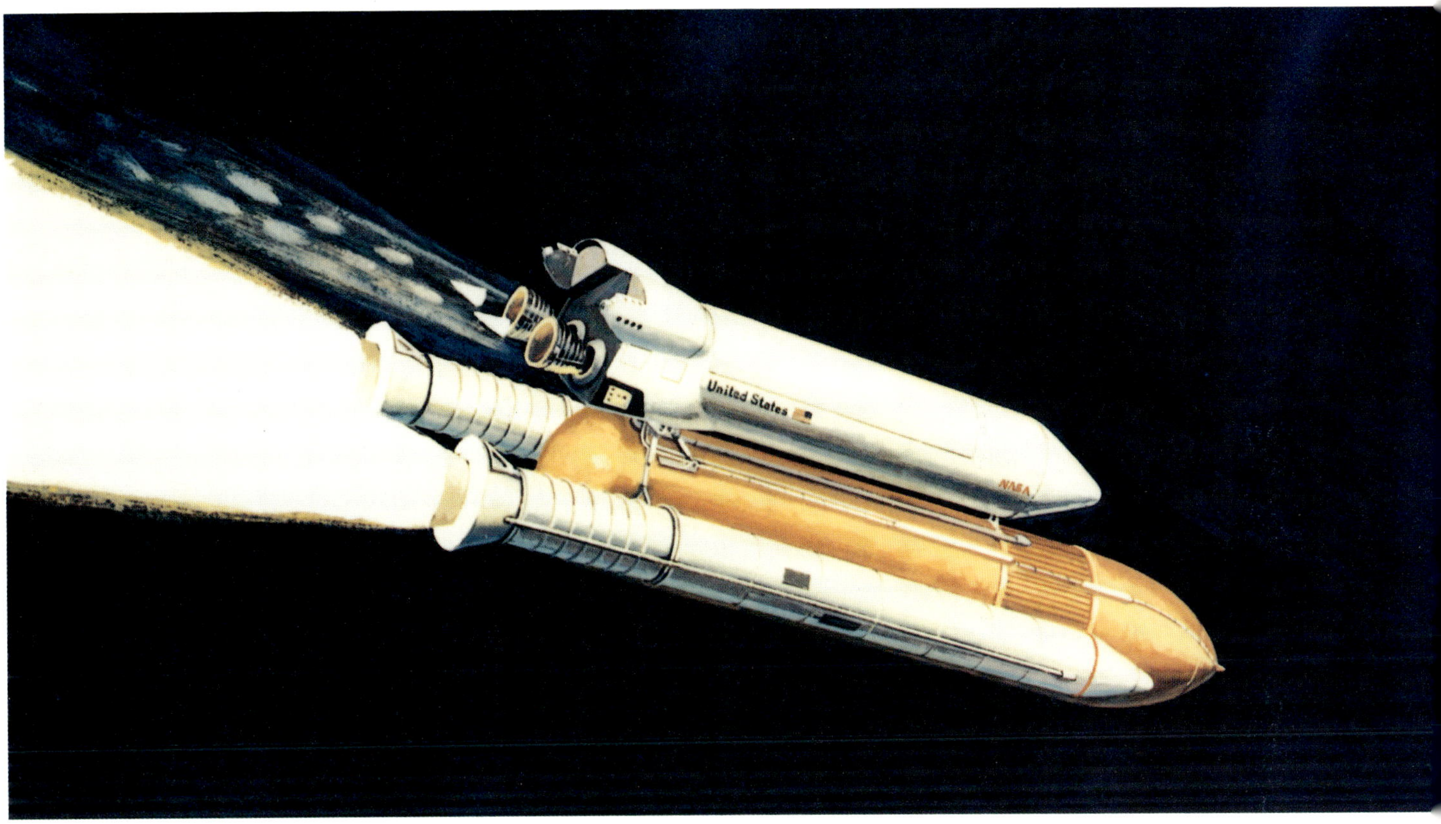

refurbishing and reusing the most costly components.

ARS has applicability not only to launch systems but to a wide variety of valuable and reusable payloads. One of the most important is the projected Crew Emergency Return Vehicle, a "lifeboat" for astronauts aboard Space Station *Freedom* in the event the Space Shuttle is not available to them.

Wind tunnel tests were conducted in the latter part of 1988 and drop tests of ARS from aircraft began early in 1989.

Orbital Manuevering Vehicle

At right, an Orbital Maneuvering Vehicle (OMV), the flat disc-like spacecraft, is being manuevered into position to rendezvous with and service an orbiting satellite. The scenario for the artist's concept in the lower photo is that a communications satellite (blue cylinder) has failed after deployment from the Shuttle Orbiter and the OMV (shown in side view at left photo) has retrieved it and is transporting it to the Orbiter for repair or return to Earth.

A reusable space tug designed to extend the reach of the Space Shuttle Orbiter, the OMV is being developed by TRW Inc.; Marshall Space Flight Center is development center; Johnson Space Center is the flight operating center; and, Kennedy Space Center is the ground operations center. The vehicle is targeted for an initial demonstration flight from the Space Shuttle *Endeavour* in 1993.

The OMV is planned initially as an adjunct to the Shuttle Orbiter, operating from the Orbiter's cargo bay and returned to Earth with the Orbiter. Later, it will be parked in space and activated for specific jobs, or it may become a payload-moving space Station adjunct.

Remotely controlled by ground-based operators employing television and other sensors to guide its movements, the versatile OMV can be configured—by use of modification kits—to perform a wide variety of space tasks, such as propelling an Orbiter-deployed payload as much as 1,200 miles beyond the Orbiter's normal operating area; assisting the Shuttle in space station assembly and construction; on-orbit servicing, maintenance or payload changeout for orbiting satellites; retrieval of spacecraft for return to Earth or reboost to another orbit; capturing and deorbiting space junk; or performing video inspections of the space station, the Shuttle Orbiter or an orbiting observatory.

Tethered Satellite

Studies have shown that tethering of two orbiting spacecraft at opposite ends of a cable offers a number of practical and scientifically interesting applications.

A tether-linked Shuttle Orbiter and satellite combination could, for example, team to tap the ionosphere's energy to generate electric power for use in space. Or a tethered satellite trailing far below an Orbiter can make scientific measurements in a little-explored region of the atmosphere above the operating altitude of aircraft and balloons but well below minimal orbital altitudes.

The environment of Space Station *Freedom* could be completely mapped by repeated "spin casting," sending an instrument payload a half-mile or so away from the manned base to acquire environmental measurements while being reeled in. Or an experiment of a possibly hazardous nature could be isolated from the station, yet a part of it, remotely positioned at the end of a tether.

These are a few of a wide range of potential applications of tethered systems that NASA has studied. Some of them will be investigated in actual flight in the 1990s. Such flights represent a complex engineering challenge. Because gravity, centrifugal force and atmospheric drag all vary with altitude, each of the two bodies in a tethered system—perhaps as much as 60 miles apart—is subject to different influences. Satellite deployment and retrieval are complicated and tether control must be precise. To demonstrate that satellites can be deployed, stabilized and retrieved on long tethers, NASA and the Italian Space Agency (ASI) National Research Council are jointly conducting a Tethered Satellite System (TSS) program.

ASI is building the TSS satellite and integrating the science payloads. NASA is developing the satellite deployment system and is responsible for payload integration and transportation to orbit; Martin Marietta Aerospace is NASA's prime contractor and Aeritalia is prime contractor for ASI. Both agencies are developing experiments to address engineering and science goals.

Three TSS missions are planned, the first targeted for a flight aboard the Space Shuttle early in 1991. TSS-1—and a subsequent TSS-3—involve experiments in electric power generation with the satellite operating 12.5 miles above the Orbiter in the ionosphere. Moving through Earth's magnetic fields, the satellite will collect electrons and send them through the conducting tether to an electron generator in the Orbiter; the generator's emission of electrons into space creates a potential difference between the Orbiter and the satellite and completes a circuit that causes current to flow in the tether. The 12.5 mile tether will generate about four kilowatts; a 60-mile tether could produce about 70 kilowatts, many times the power level of the Shuttle Orbiter. Thus a tethered system could be employed as a space power generator, supplying power to spacecraft to supplement other power sources or serving as an emergency backup.

On the TSS-2 flight, the satellite will be deployed downward on a 60-mile tether to examine the lower thermosphere and make atmospheric science measurements.

Space Station Freedom

At right is an artist's concept of the baseline configuration of Space Station *Freedom*, which will be permanently occupied by U.S./international astronauts for 30 years or more beginning in 1996. In addition to the manned base, the space station complex includes a number of independently orbiting experiment platforms and laboratory-type spacecraft; initially there will be two polar orbiting platforms, one supplied by the U.S. (lower photo), the other by the European Space Agency (ESA). The other participating nations are Canada and Japan.

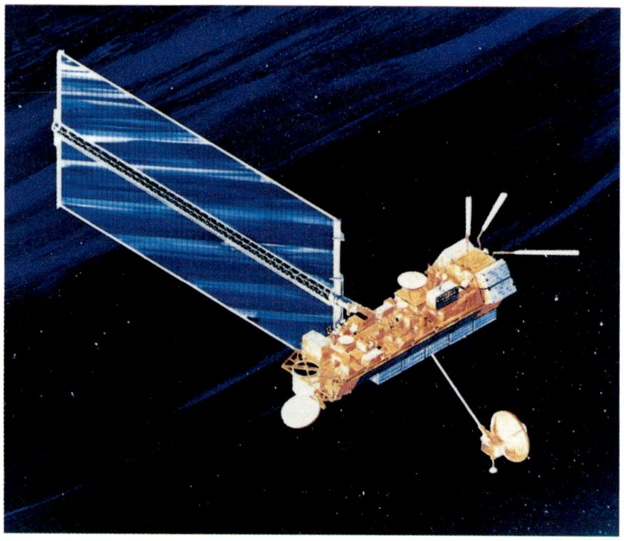

The manned base is built around an integrated truss assembly that provides the structure for the four pressurized modules and the supporting elements and systems. The main transverse boom, measuring 508 feet in length, suspends eight solar array "wings," each wing containing 14,592 solar cells to supply power to the station. The truss has provisions for attaching, in addition to the pressurized modules, logistics carriers, externally-mounted experiments, and instruments for viewing Earth and the universe.

In the spring of 1989, Goddard Space Flight Center awarded a contract to Martin Marietta Space Systems for development of the Flight Telerobotic Servicer (FTS), one of the last major space station systems to be assigned for development. The FTS is a space robot that will play a role in the assembly of Space Station *Freedom* and later will become a station adjunct handling servicing and repair tasks.

At upper right, the FTS is using both its highly dextrous robotic arms to repair a station component. Controlled by an astronaut at a workstation in the Shuttle Orbiter or the space station, the FTS has video cameras for "eyes" that enable the controller to see from the robot's perspective; for assembly or servicing jobs, the robot's motions are directed by advanced artificial intelligence computers. The FTS will employ robotics and machine intelligence to maximum extent, to provide a focus for advancement of these technologies in the interest of U.S. competitiveness.

Work Package 3, under Goddard and its prime contractor—General Electric Astro Space—will provide accommodations for attached payloads, and the U.S. unmanned polar orbiting platform, which will be used initially by NASA's Earth Observing System (EOS) program. Managed by Goddard, the EOS program is a comprehensive effort to obtain a scientific understanding of the entire Earth system on a global scale by determining how Earth's component parts and their interactions have evolved, how they function and how they may be expected to evolve in the future.

Other space station development and fabrication activities include:

Work Package One, managed by Marshall Space Flight Center with Boeing Aerospace Company as prime contractor, involves provision of the U.S. laboratory and habitation modules; logistics elements, resource node structures, the environmental control and life support system, internal audio and video systems and associated software.

Work Package Two, Johnson Space Center and McDonnell Douglas Astronautics Company: truss structure, airlocks, outfitting the resource nodes, data management software and hardware, the communications and tracking system, the guidance, navigation and control system, extravehicular activity systems, the propulsion system, thermal control system and associated software. This package also includes the U.S. contribution to the Mobile Servicing System (MSS) shown above; the U.S. assignment is a rail-mounted mobile transporter that will move along the truss carrying the Canadian-built Mobile Servicing Center. The MSS will play the main role in *Freedom* assembly and

maintenance, moving equipment and supplies around the station, releasing and capturing satellites, supporting astronaut extravehicular activities, servicing attached payloads, and loading/unloading materials from docked Shuttle Orbiters.

Work Package Four, Lewis Research Center and Rocketdyne Division of Rockwell International: the complete power system and associated software.

(Continued)

Space Station Freedom

(Continued)

Located in the center of the transverse boom of Space Station *Freedom* are the pressurized living and working areas, a U.S.-built habitation module and three laboratory modules, one each to be provided by the U.S., ESA and Japan. Linking the modules are four pressurized "resource nodes" that serve as command and control centers and as passageways to and from the various modules.

At right is a concept of the two U.S. modules, the habitat module at top center and the laboratory module below it, with resource nodes at either end and, at far left, a logistics module for transport and storage of supplies and experiments. Two of the four resource nodes have "cupolas," one an observation post looking toward Earth (right below), the other facing outward to space.

ESA's *Columbus* laboratory is of similar size and shape to the U.S. lab module, and it also has an airlock for temporary exposure of experiments to the vacuum of space or for transfer of experiments to support external activities. The Japanese Experiment Module (top center) includes a short pressurized segment to which a dome-shaped experiment logistics module can be attached, plus an "outdoors" facility for experiments wherein unpressurized space conditions are preferred or essential.

The habitation module, designed for a maximum of eight astronauts, has everything needed for long duration occupancy, including facilities for eating, sleeping, hygiene, work activities and health care. At far right top is the Crew Health Care System, which includes exercise equipment, a diagnostic laboratory, therapeutic equipment and medications for treatment of minor illnesses, a computerized data base of medical information, and an imaging system downlinked to a network of consulting physicians on Earth.

One concept of the habitation module's

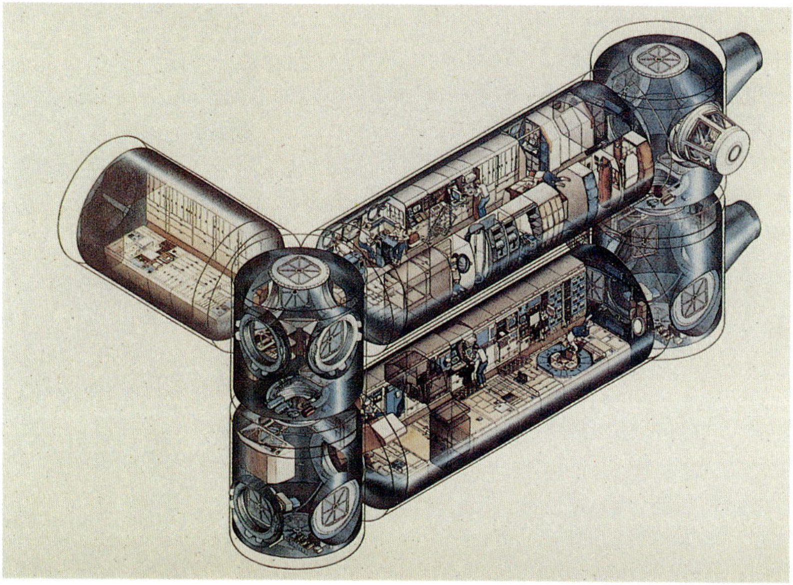

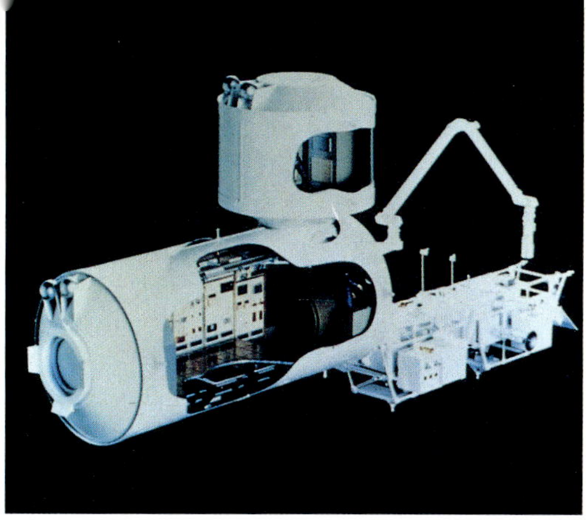

interior is shown at middle right. The crew member in center photo sits at a workstation while another, in the background, prepares a meal at the module's galley; not visible is a wardroom opposite the galley for dining, conferences or audio-video entertainment.

At mid-module are the personal hygiene facilities and at the far end from the galley are private compartments for each of the eight crew members. Each compartment has 150 cubic feet of space that includes a TV, a portable workstation, stowage for clothing and personal effects, and a "vertical bed," a sleeping bag hung on a wall; at lower right, one crew member is tuning in the TV while another sleeps.

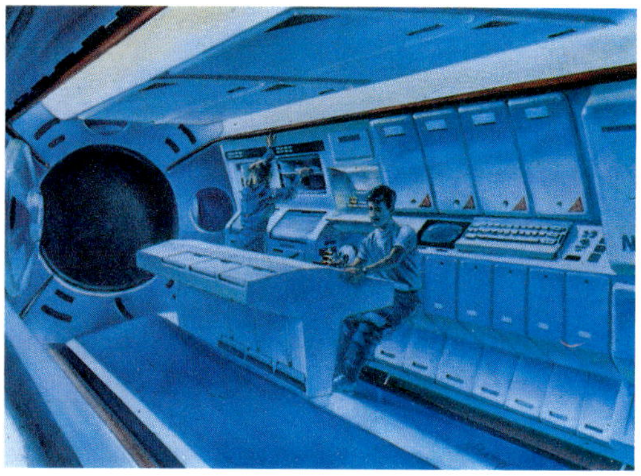

Assembly of Space Station *Freedom* will require approximately 20 Shuttle flights. There are four major milestones in the assembly sequence. The first, known as the First Element Launch, involves delivery—in the first quarter of 1995—of a "cornerstone" set of components for the manned base, including a power module, communications equipment, a reaction control system, the mobile transporter, an assembly work platform and parts of the truss. The second milestone occurs on the fourth flight, tentatively in the fourth quarter of 1995, when the U.S. laboratory has been placed into service and components have been added that will provide the base with an interim mantended capability. The third milestone is permanent manned capability, targeted for the fourth quarter of 1996. The final milestone is "assembly completed," with all modules and components in place and an international crew on board; the target date as of midsummer 1989, subject to revision, is the first quarter of 1998.

TECHNOLOGY TWICE USED

A representative selection of new products and processes adapted from technology originally developed for NASA mainline programs, underlining the broad diversity of spinoff applications and the social/economic benefits they provide

Spinoff developments highlighted in this section are based on information provided by secondary users of aerospace technology, individuals and manufacturers who acknowledge that aerospace technology contributed wholly or in part to development of the product or process described. Publication herein does not constitute NASA endorsement of the product or process, nor confirmation of manufacturers' performance claims related to the particular spinoff development.

Spinoff From A Moonsuit

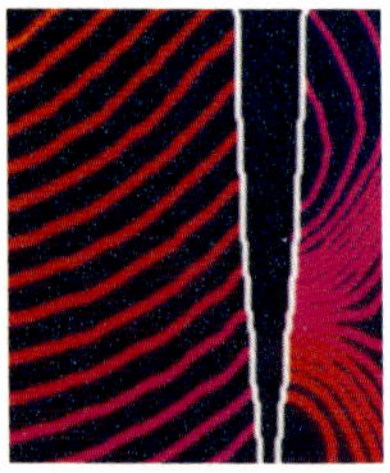

Stevie Roper of Waynesville, North Carolina, age 11, can now ride a bike, play baseball, go to the zoo or help his father rake the leaves. So what's exciting about that, you ask.

To Stevie, all of it. Not so long ago he couldn't do any of those things, nor engage in any kind of activity that would cause his temperature to rise. He faced the constant threat of heat exhaustion or stroke because he suffers from what he calls "hypo-whatever." It's actually hypohidrotic ectodermal dysplasia, HED for short, and it means that Stevie was born without sweat glands, which allow body heat to escape and thus serve as a natural cooling system for normal people.

A space-derived personal cooling system exemplifies the benefit potential of aerospace technology transfer

For the first nine years of his life, Stevie couldn't even venture outdoors in summer. He had to spend hours daily bathing or getting "greased." At school he frequently had to wet himself down fully clothed. He suffered from a variety of HED-related afflictions and was often hospitalized.

In 1987, Stevie had a close call, a severe heat stress situation of critical dimension—but in the long run that turned out to be the best thing that ever happened to him. It triggered a series of events that resulted in relief for Stevie and others, perhaps eventually for all of the estimated 400 to 1,000 HED people in the United States and many more elsewhere. Thanks to NASA space suit technology, the cooperative "can do" spirit of a cooling system manufacturer, and the relentless perseverance and dedication of Stevie's aunt, Sara Ann "Tootsie" Moody of Hampton, Virginia.

The story begins on a summer eve in 1987 when Stevie, visiting Hampton, took a short drive with Sara Moody's daughter. Because of Stevie's HED problem, the trip had been cautiously scheduled for after sundown. Even so, Stevie began to pant in the non-airconditioned car and within minutes was overheated, close to collapse. Quick thinking by the daughter, who swerved to the curb, borrowed a lawn hose and wet him down, may have saved Stevie's life.

That incident frightened Tootsie Moody and set her in action. She telephoned NASA's Langley Research Center in Hampton and suggested that the organization that had sent men to the moon surely must have developed a technology that could help Stevie Roper. A NASA official listened to Sara's story, did some research and reported back that there was indeed a spinoff product that might alleviate Stevie's problem. He referred her to Life Support Systems Inc. (LSSI), Mountain View, California, manufacturer of Micro Climate personal cooling systems for people in occupations where elevated body temperatures threaten fatigue and collapse.

(Continued)

Photo by John H. Sheally II

Stevie Roper (top), who was born without sweat glands, lives a closer-to-normal life with his space spinoff "cool suit," which prevents overheating. Coolant circulates through tubes in the torso vest and headpiece (under the cap) he is wearing; his mother is connecting the suit's hose to a cooling/pumping pack.

At right is Sara Ann Moody, a Hampton (Virginia) housewife who is spearheading a program to provide cooling garments for children with disorders that cause life threatening overheating. With her is Roger Bear, an official of the Po Folks restaurant chain, who helped Sara launch a successful fund raising drive.

Spinoff From A Moonsuit

(Continued)

LSSI is an aerospace spinoff company, one commercialized to promote Earth uses of technology originally developed for aerospace programs. The Micro Climate technology had its origin in a 1968 NASA development program at Ames Research Center that produced a spacesuit undergarment for cooling astronauts on the surface of the moon or during forays outside their spacecraft or space station; the system circulated a fluid, cooled by a heat exchanger and delivered by a battery-powered minipump, through a network of tubes in the garment.

In 1971, Ames awarded a contract to Acurex Corporation for an extension of the technology involving development of a heat stress alleviating liquid-cooled helmet liner for helicopter pilots. In the mid-1970s, NASA and the Bureau of Mines jointly sponsored an Acurex program for development of a self-contained cooling system for mine rescue work. In 1980, William Elkins, formerly with Acurex and long associated with cooling system research, founded LSSI to pursue commercial uses of the technology. LSSI has refined the technology and brought to the commercial marketplace three generations of improvements.

Gary Rodne, LSSI's director of commercial operations, took the call from Sara Ann Moody. Yes, he said, after some reflection, LSSI could adapt its cool suit technology to Stevie's needs; it would cost about $4,000. That was a big barrier; Stevie's parents couldn't afford it, nor could Tootsie Moody.

With the same dogged determination that had started her on her quest, Sara Ann Moody tackled the money problem. She got a local printer to supply—free—posters with Stevie's picture and his story. With the help of friend Roger Bear, district manager for Po Folks Restaurant Corporation, she placed posters and coin collection jars in several Po Folks restaurants. Another friend, Chesapeake (Virginia) businessman Tom E. Arney, started a separate fund raiser for Stevie. Within five weeks, Sara had raised $5,000. LSSI, meanwhile, had decided to charge only for the suit's materials—$2,600. The extra money was applied to Stevie's extensive medical bills.

In October 1987, Stevie Roper was presented his cool suit, a miniature version of LSSI's Mark VII Micro Climate System®, which uses a rechargeable battery and a small pump to circulate an ice-pack-chilled fluid through a tube-lined headcap and wraparound torso vest. The system is designed to eliminate 40-60 percent of Stevie's body heat and lower his heart rate by 50-80 beats a minute.

Stevie Roper improved rapidly after he started using the cooling garment. "He became a different person," says Aunt Sara, "no longer

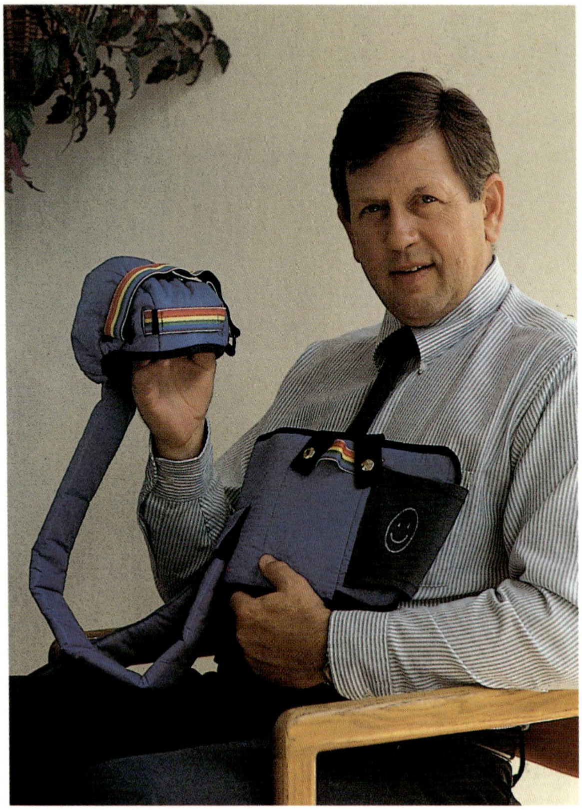

Gary Rodne of Life Support Systems Inc. displays a HED cool suit—headcap and torso vest—that is a child-size version of the company's MicroClimate line of protective garments for workers whose jobs subject them to heat stress.

®Mark VII Micro Climate System is a registered trademark of Life Support Systems Inc.

withdrawn, no longer frail but filling out—it's as if he'd been in a closet, afraid to come out, but now he's very happy."

Sara Moody was happy, too, for Stevie and his parents, and because she felt she had spearheaded something important. LSSI president William Elkins and marketer Gary Rodne were happy that their company had been able to perform an outstanding service. Presentation of the "Stevie Suit" could have been a happy ending to the story. As things turned out, it was just a beginning.

When Stevie Roper received his cool suit, it was something of a media event. The story was carried by newspapers and television from coast to coast and a documentary film was shown on about 800 TV stations in the U.S. and abroad.

Then Tootsie Moody's telephone began to ring. There were calls from all over the country from families with the HED problem or related diseases. And there were messages from abroad—from India, New Zealand, Kuwait, Saudi Arabia, Argentina. There was a common theme to the comments: How can I get a cool suit? Why don't you start a foundation to help all the children who need such suits?

"Well, why not a foundation?" said Sara Ann Moody, who proceeded to attack this new assignment with characteristic vigor and determination. Within a few months she had established the HED Foundation, Hampton, Virginia 23[illegible]0. Joining her on the board were Po Folks' Roger Bear and businessman Tom Arney who had helped her launch the Stevie Suit campaign, Stevie's mother, NASA Langley official Keith Henry, Tootsie's husband, a Hampton doctor and lawyer, and the man at the other end of the lifeline, LSSI's Gary Rodne.

With Roger Bear's help, Tootsie Moody extended her contribution base to embrace 166 Po Folks restaurants all over the country; each displays a HED poster and a king size pickle jar for donations. In mid-1989 the HED Foundation's income was running about $4,000 a month from the Po Folks jars alone and there were occasional outside contributions. At publication time, the foundation had raised well over $40,000.

The money was put to good use. Six months after Stevie Roper pioneered the movement, five-year-old Scott Gibson of Trussville, Alabama became the second recipient of a cool suit. July 29, 1988 was a big day, a double suit presentation to a pair of six-year-olds, Nick Elmore of Knoxville, Tennessee and Ossama Abdulla of Kuwait, the first foreign suit recipient. By the end of 1988, eight children were wearing cool suits, including the second foreign youngster, Tarunpreet Arora of Jalandhar, India. *(Continued)*

An LSSI engineer is using specifications of an adult-size vest as a basis for designing the scaled-down HED vest.

Spinoff From A Moonsuit
(Continued)

This year, with a greater pickle jar yield, the pace of suit deliveries has picked up; on one occasion there were four presentations in a single month. At publication time, about 20 suits were in use, but the word was spreading and new applicants were swelling the waiting list. The HED Foundation's backlog had not diminished appreciably; there were another two dozen in line.

LSSI was still providing the suits at near cost. Company officials were delighted with the success of the program and they had no trouble coping with the increasing HED workload because it represents a small part of LSSI's operation.

Since its founding in 1980, LSSI has grown into a thriving business that has expanded both horizontally—more and more applications—and vertically—increasing orders for some of the principal applications. Micro Climate cooling systems are in service with U.S. and foreign military forces for use by personnel who must perform arduous tasks while wearing hot and bulky protective gear, for airmen flying unpressurized aircraft, for armored vehicle crews, and for shipboard personnel engaged in heat stressful work, such as boiler room or steam catapult room operations.

The range of civil applications is even broader. It includes protection for public service and industrial firefighters and hazardous materials controllers, plus workers in such industries as nuclear power, primary metals reduction, glass manufacturing, chemical processing, petrochemical refining, paper production, steel mills and foundries, and agricultural crop dusting.

LSSI is meeting a special sports need by providing—through Carlson Technology Inc., Livonia, Michigan—cooling equipment for professional race car drivers. Engine heat and aerodynamic measures that detour cooling air away from the cockpit create cockpit temperatures of 130-140 degrees Fahrenheit and such heat, sustained for long periods, can cause fatigue, dehydration, even collapse. So more and more drivers are turning to cooling suits. LSSI's list of clients includes such well known names as Richard Petty, A.J. Foyt, Dale Earnhardt, Ayrton Senna, Bill Elliott and Al Unser Jr., plus about 250 others.

A new sports application made its appearance in 1988. The ThermoAire Splint™ provides pneumatic compression with integrated cooling to replace ice packs and elastic bandages for sports injuries. The compression reduces fluid swelling while the cooling reduces the level of injury. LSSI's ThermoAire Splint™ made news in 1988 when San Francisco 49er quarterback Joe Montana tried one on a knee injury and was able to play the following week.

Scott Gibson of Trussville, Alabama, five years old when he received his cool suit, was the second recipient and the first after the establishment of the HED Foundation.

™ ThermoAire Splint is a trademark of Life Support Systems, Inc.

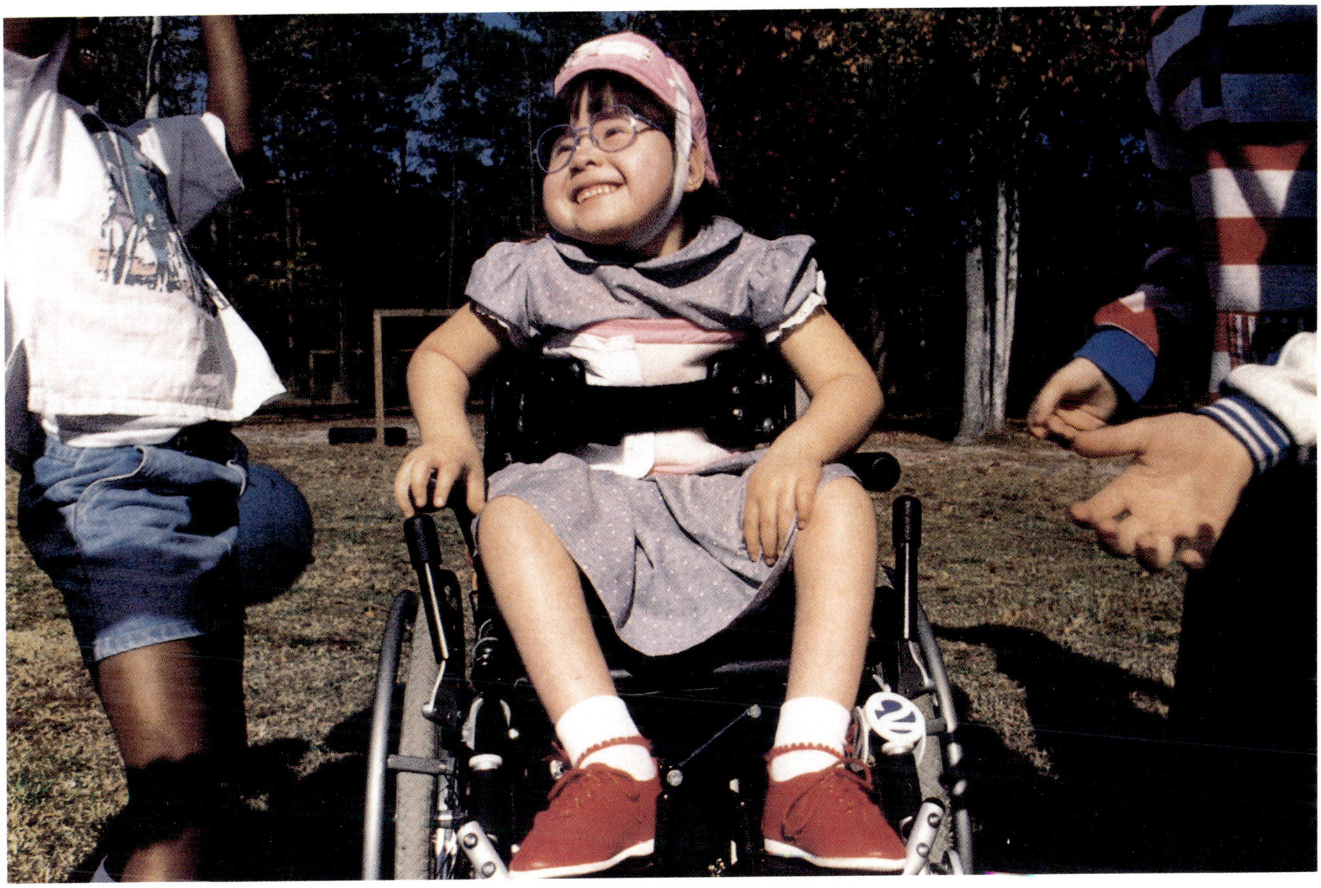

Lynette Bowers of Myrtle Beach, South Carolina models her lavender cool suit. Many of the suits require special design to meet the particular needs of the wearer.

LSSI, in conjunction with the Jimmy Heuga Center and the University of Utah Medical Center, recently completed a pilot research program that clearly demonstrated significant benefits to multiple sclerosis patients, who must exercise but often cannot tolerate heat build-up caused by exercise. In the same program very significant results were also observed for a patient with limb neuropathy, a very painful disorder. In a separate experiment, conducted by Dr. Arnold Malcom at St. Joseph's Medical Center, Burbank, California, LSSI's vest was used with positive results for heat therapy of extensive superficial cancer. Additionally, the NASA-sponsored Research Triangle Institute Applications Team, Research Triangle Park, North Carolina is working toward commercialization of a cooling garment specially designed for use by quadriplegics, who are often unable to tolerate heat stress because they are unable to perspire below the level of injury to the spinal cord.

(Continued)

Spinoff From A Moonsuit

(Continued)

Tootsie Moody's phone is still ringing. With some four dozen cooling suits already delivered to HED youngsters or on the agenda, she feels the HED Foundation's work will continue for a long time because she is continually hearing from new clients who have just learned of the foundation's existence.

And her horizons are broadening. Sara Moody, who not long ago couldn't spell hypohidrotic ectodermal dysplasia, can tick off a list of related afflictions whose victims are prone to overheating—lamellar ichthyosis, Cursayer Syndrome, cystic fibrosis, multiple sclerosis, severe burns and many other disorders. She knows about them because the victims' families, desperate for a shred of hope, have turned to her. She refers them to LSSI, but it pains her deeply that she can't do more, because the HED Foundation hasn't yet enough income to expand beyond its chartered purpose.

"Not yet," says Tootsie Moody. "But it's there in my mind. I want to help *all* the children. If we can increase the contributions . . ." She's off on a new dream and when Sara Ann Moody gets to dreaming, things start to happen.

And what's in it for Tootsie? She has a ready answer.

"A wonderful satisfaction. I've never enjoyed anything as much as the activity of the last two years. It's so rewarding to see the faces of the kids and their parents when they first try the cool suit and know it works.

"And later they respond to questionnaires and send me notes. It's so exciting to read them, I get chills down the spine. It's not just a suit we're giving these kids—it's new hope."

The cool suit story is a classic example of the spinoff potential for both economic gain and public benefit. Spinoffs often produce product sales in the multimillions of dollars annually. In other instances, spinoffs offer only moderate economic gain but provide public benefit in other ways, ranging from simple conveniences to significant developments in medical and industrial technology.

LSSI represents one side of the spinoff coin; here a technology transfer resulted in establishment of a prospering company, with attendant benefit to the Gross National Product and job creation.

Tootsie Moody's HED Foundation is the other side of the coin, a compassionate movement made possible by space technology, a program that is already helping many, providing new hope for many more, and offering great potential as inspiration and catalyst for further spread of the medical applications of body cooling.

A particularly grateful cool suit recipient is Krystal "Toby" Sharrett, 15, of Atlanta, Georgia, whose lack of sweat glands in the feet caused serious sores and threatened amputation. Presented a cool suit in April 1989, Toby improved dramatically and by June her feet had completely healed.

For Toby, LSSI designed a special unit with cooling slippers, shown being fabricated at left; at right is a commercially-available slipper.

For the past 27 years, under its Technology Utilization Program, NASA has been actively engaged in encouraging the secondary application of aerospace technology. During that time upwards of 30,000 aerospace originated innovations have found their way into everyday use. Collectively, these spinoffs represent a substantial return on the aerospace research investment in terms of economic gain, lifestyle enhancement and solutions to problems of public concern.

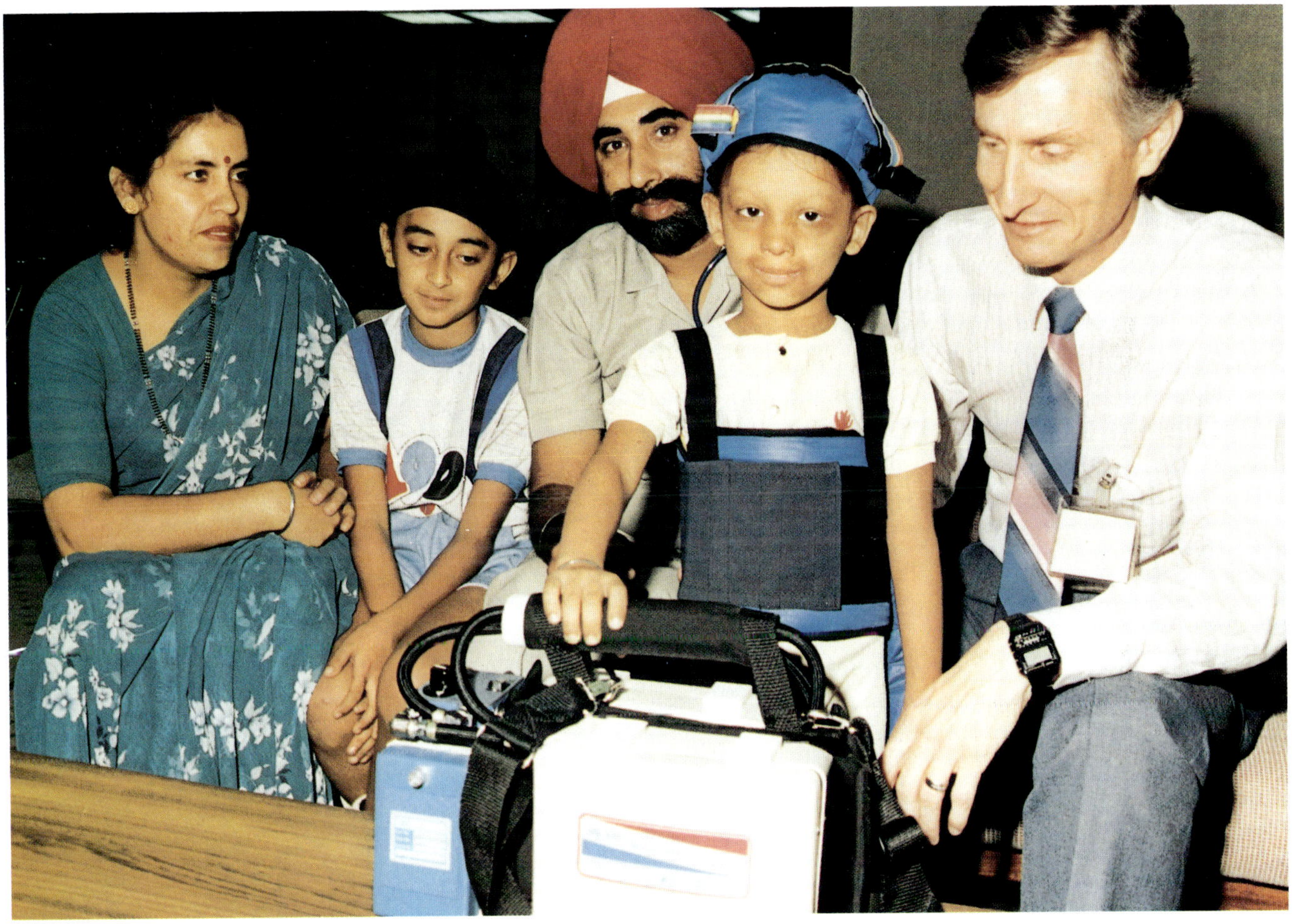

The Arora family of Jaladhar, India pose with HED representative Keith Henry (right), who presented a cool suit to four-year-old Tarunpreet Arora. In foreground is the suit's cooling pack and pump, which circulates a chilled fluid through the tube-lined headcap and vest.

Cropduster pilot Gary Owens (left) solved the problem of high cockpit temperatures with his LSSI cool suit. So did NASCAR Winston Cup champion Dale Earnhardt (right) one of more than 250 race car drivers who wear LSSI equipment.

Breaking the Ice

Ice buildup on the wings of aircraft is a serious problem, a potential trouble source that can cause accidents or damage to the airplane. Which is why the military services, airline operators, aircraft manufacturers and public safety agencies are expressing great interest in a new NASA development, a deicing system that not only promises improvement in flight safety but offers bonuses in airplane performance and economics.

A supereffective ice removal system for use on aircraft, ships or bridges tops a selection of spinoffs in public safety

The system is known as the Electro-Expulsive Separation System (EESS) and it won for its inventor, Ames Research Center senior engineer Leonard A. Haslim, the 1988 NASA Inventor of the Year Award.

EESS uses only one thousandth the power of existing electro-thermal deicers and weighs a tenth as much. It can be used on helicopter rotors and jet engine inlet ducts as well as aircraft wings; it can be installed on new aircraft or retrofitted to aircraft already in service.

EESS is now commercially available. In November 1988, following a series of successful evaluation tests, including wind tunnel and flight testing at Lewis Research Center, NASA awarded a patent license to Dataproducts New England, Inc. (DNE), Wallingford, Connecticut. This first license grants DNE exclusive rights to market EESS for large turbine-powered airplanes and helicopters. NASA may award other licenses for small turbine-powered aircraft and for large and small propeller-driven aircraft. DNE has initiated a joint venture with American Airlines to study the specifics of an EESS for the airline's entire fleet; they are joined by McDonnell Douglas Aircraft Corporation in an examination of EESS installation in the company's airliners.

EESS consists of an elastic, rubberlike deicer boot on the wing leading edge, with flexible conducting copper ribbons embedded in the boot. The conductors are separated by slits in between and parallel to the ribbons. When the system is switched on, a bank of capacitors in the power supply discharges into the conductors. The strong direct current pulse, discharged in less than a millisecond, suddenly induces the conductor pairs to repel one another. The result is a violent and powerful force that causes the slit-voids to expand explosively; ice on the wing is instantly pulverized and ejected.

This technique overcomes many of the limitations of other deicers. For example, jet transport aircraft generally use systems that bleed hot air from the engines to melt wing ice, but the bleed reduces engine thrust and increases fuel consumption; use of EESS costs no penalty in performance or economics. Pneumatic deicing boots operate slowly and will not break ice until it becomes one-quarter to one-half inch thick; when it does, big chunks may damage engines or airfoils. EESS has demonstrated ability to eject ice of any thickness from mere frost to an inch-thick glaze; it does so instantly and pulverizes the ice so that no shard is large enough to damage the plane.

EESS offers similar advantages over all existing systems; generally it is more flexible, more effective in cracking and ejecting ice, smaller in both weight and space requirements, easier to maintain and repairable in the field.

Although EESS was developed as an aircraft system, it has other important applicability — for instance, in the problem of ship icing. Shipbuilders and ship operators are interested in an effective way to remove ice from ship superstructures. A big ship like a guided missile cruiser can accumulate very large ice accretions that could impact performance. Ingalls Shipbuilding, builder of U.S. Navy Ships, is participating in development and test of a shipboard EESS. Westinghouse Electric Corporation is developing a ship deicing system and DNE is also investing ship applications.

A ship deicing system would be a major boon to the nation's fishing fleets, such as those operating off the coast of Alaska, where harsh icing conditions are a potential hazard. Ames Research Center is working with the Alaskan Department of Commerce and Economic Development to explore the possibilities of installing a maritime EESS on Alaskan fishing vessels.

EESS also shows promise for bridge deicing. Bridge icing is a hazard to drivers and a dangerous, costly threat to the bridge's structure because chemical removal of the ice layer can cause structural and surface damage. The Federal Highway Administration has enlisted NASA's aid toward solution of the problem.

And it doesn't stop there. Haslim sees applications of the electro-expulsive concept in industry, for example in the bakery and plastics industries as part of a mold release mechanism to speed up production lines. And it might even be applied to a synthetic self-pumping artery for heart patients. Haslim is working with a Stanford University medical engineer on that application. Although the research will take many years, he thinks "We can make it happen.

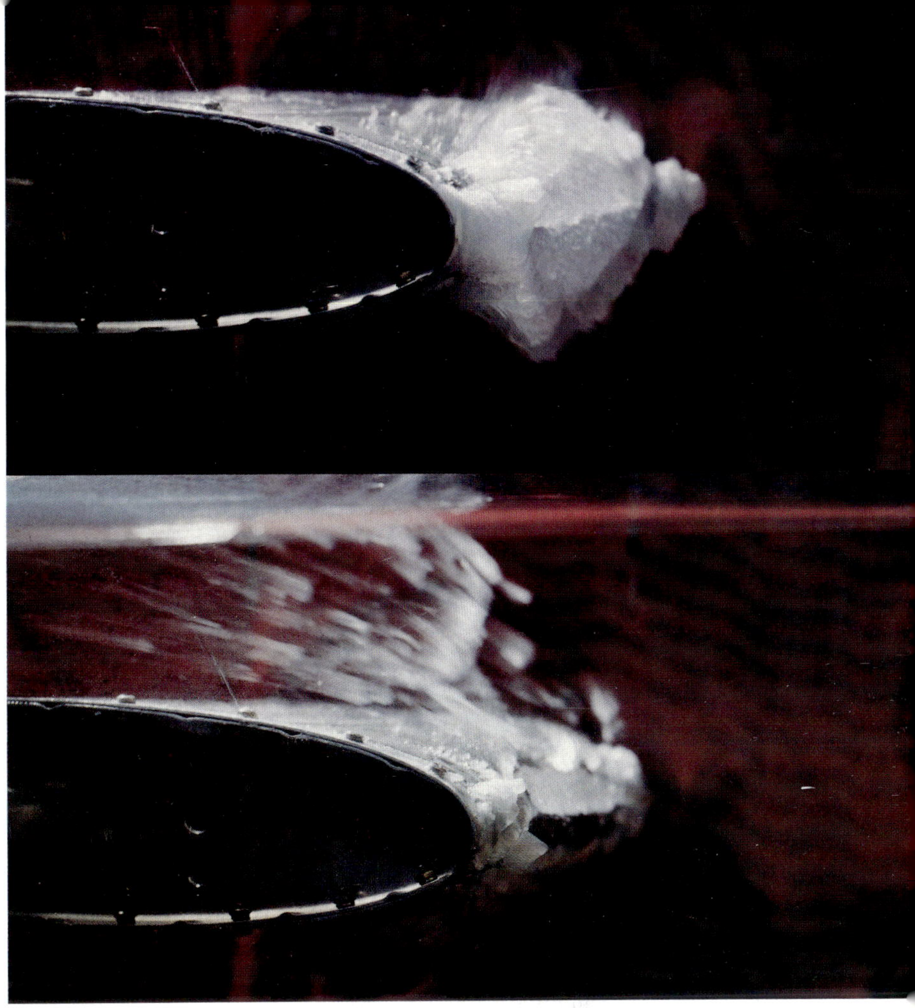

These sequential photos show a wind tunnel test of an innovative Electro-Expulsive Separation System (EESS) on an ice-covered aircraft wing segment. In the top photo, EESS has just been activated; in the middle photo, taken less than a millisecond later, most of the ice is gone, pulverized and ejected by an EESS-created electromagnetic force. In the lower photo is Ames Research Center senior engineer Leonard A. Haslim who won the Inventor of the Year Award for developing EESS, which offers a wide range of advantages over existing aircraft deicing systems.

Airline Crew Training

"Until the late 1970s, airline crew training was focused on stick-and-rudder skills, the physical skills needed to fly an airplane," says United Airlines training coordinator Cliff Lawson. "Training in crew coordination or teamwork skills was almost unheard of. By the late seventies, however, the airline industry and United in particular began to realize that the vast majority of accidents and incidents were taking place in airplanes that were flyable. In many cases, there was absolutely nothing wrong with the airplane."

What was wrong was human error, surveys showed; statistics indicated that mistakes were causing far more accidents/incidents than mechanical malfunctions. Among the most common human errors, researchers found, was failure on the part of crew members to communicate properly and failure to use resources readily available to them. These and many other breakdowns of cockpit management and teamwork sparked a major change in airline flight and simulator training wherein the focus shifted to programs intended to improve teamwork and coordination in the cockpit.

Ames Research Center, a pioneer and world leader in human factors research aimed at greater understanding of aircrew capabilities and limitations, played a principal role in developing two interdependent training programs that concentrate on more effective aircrew operations.

The first is Cockpit Resource Management, or CRM, which by one definition is "the effective utilization of all available resources — hardware, software and liveware — to achieve safe, efficient flight operations." Hardware includes the aircraft's on-board systems; software embraces operating manuals, bulletins, charts and other information; liveware is people, including air traffic controllers and other "resources" not on board. CRM defines a number of areas for focus in lectures and simulation training, such as decision making, establishment of priorities for cockpit attention, cockpit workload distribution, making most effective use of available information, avoiding distractions, developing intracockpit communication skills and, for the pilot in command, establishing firm leadership that nonetheless encourages participation and critique by subordinates.

Ames' Aerospace Human Factors Research Division began CRM development in the latter 1970s and — in 1979 — sponsored the first CRM workshop, which brought together airline and government aviation interests to address the problems of human error.

A major result of that workshop was United Airlines' initiation, with NASA help, of a multifaceted CRM training program. Many other carriers adopted the United program and still others developed their own approaches, again with NASA assistance. A typical comment, from Captain Reuben Black, chairman of Delta Airlines' CRM Steering Committee: "NASA has been a tremendous help to us...The staff of Ames Research Center have been very influential in the development of our CRM program." In top left photo is a Delta CRM session.

Ames also played a leadership role in the development and airline adoption of a CRM-related program known as LOFT, for Line Oriented Flight Training, in which crews fly complete missions in high fidelity simulators that simulate "real world" conditions as closely as possible. In photo far left is United Airlines' LOFT simulator; at left is a view from the simulator cockpit of a realistic final approach.

In LOFT training, instructors induce emergency situations whose successful handling demands coordinated actions of all crew members. The entire simulated flight, conducted in real time, is video-taped for later review; at right, a United captain and first officer (seated) go over their tape with their simulator check pilot. "This self critique is one of the most valuable facets of CRM training," says United Airlines' Lawson. "It allows an individual pilot to see himself as others see him and consequently to observe how his individual operating style affects other crew members."

"LOFT is a learning experience in which errors are usually made," according to two of the leading CRM researchers, Dr. H. Clayton Foushee, formerly of NASA-Ames, and NASA consultant Dr. Robert L. Helmreich of The University of Texas at Austin, in their chapter of the book Human Factors in Aviation.© "However, since effective group function in this environment is by definition the management of human error, LOFT provides highly effective crew coordination training."

NASA continues to conduct or support research to improve CRM/LOFT training. In 1986, Ames — jointly with the USAF Military Airlift Command — sponsored a second CRM workshop. Under a NASA grant, The University of Texas at Austin, with participation by several airlines and some military units, is collecting data on CRM/LOFT; the intent is to help users measure the before-and-after effects of resource management training as an aid to enhancing their CRM programs.

Anti-glare Filters

It is estimated that there are more than 15 million cathode ray tube (CRT) information systems operating in the United States and the number is growing rapidly. Widespread use of CRTs poses an occupational hazard for some.

In a survey of CRT operators conducted by the National Institute for Occupational Safety, 95 percent of those interviewed complained about screen glare and 65 percent cited specific problems: blurred vision, itching or burning eyes, eyestrain, fatigue and headaches.

To guard against loss of operator productivity, computer manufacturers and CRT-using businesses are employing a variety of anti-glare devices and techniques. A particularly effective technique is use of a patented coating known as HEA® (for high efficiency antireflection), developed by Optical Coating Laboratory Inc.(OCLI), Santa Rosa, California in the early 1960s as a means of improving visible light transmission in aerospace vehicles. It was used to coat the windows of the Gemini two-man spacecraft in 1963 and it has been employed on all subsequent manned spacecraft from Apollo to the Space Shuttle.

OCLI now sells HEA-coated panels to many original equipment manufacturers in the computer industry, who bond the panels directly to their CRTs. Additionally, OCLI offers a line of retrofit products, anti-glare filters known as Glare/Guard®, which employ the same thin film coating technology. The coating minimizes the reflected brightness to which the human eye is most sensitive; in addition to significant glare reduction, it provides enhanced image-to-background contrast, high resolution and improved readability. In an evaluation of 10 types of retrofit anti-glare filters by Virginia Polytechnic Institute and State University, a team of visual scientists judged that Glare/Guard filters provided the best display image.

At left, a HEA-coated demonstration disc shows how the filter eliminates glare from the Compaq SLT 286 laptop computer. At right top, the glare is evident in an unfiltered CRT; the model is simulating the associated eyestrain. In the middle photo she is illustrating a solution: use of a Glare/Guard filter, which can be installed in minutes. The bottom photo shows CRTs equipped with two different types of OCLI filters, the Vantage™ on the left and the Professional Plus™.

Produced in 28 sizes to fit more than 2,000 types of monitors, Glare/Guard filters are OCLI's first consumer products, but for more than 40 years the company has been a leading producer of thin film coated products for defense, space, industrial and scientific applications. In addition to the Glare Guard line, OCLI sells HEA under the name Invisiglass®, used in multielement optical systems; readout panels for medical diagnostic and test equipment such as x-ray machines, heart rate monitors and breath analyzers; oscilloscope windows for scientific test equipment and portable analyzers; and energy-efficient lens systems for projectors, copiers and large screen video systems.

®HEA, Glare/Guard and Invisiglass are registered trademarks of Optical Coating Laboratory Inc.

™Vantage and Professional Plus are trademarks of Optical Coating Laboratory Inc.

Radiation Blocking Lenses

At right is a selection of Suntiger® sunlight-filtering glasses that protect human vision by blocking blue, violet and ultraviolet light which, occupational safety research has shown, can cause a variety of eye disorders, in particular cataracts and age-related macular degeneration.

Produced by Suntiger Inc. Biomedical Optics, North Hollywood, California, Suntiger PST™ (Polarized Selective Transmission) lenses bar 99 percent of the potentially harmful wavelengths while allowing the visually useful colors of light (red, orange, green) to pass through. Similar in principle to the natural filters in the eyes of hawks and eagles, they also block out high intensity glints of reflected sunlight that cause localized areas of the retina to receive high doses of light, or glare. Additionally, the lenses improve visual acuity, night vision and visibility through haze, fog or smog.

Introduced in the early 1980s, the PST lens was a spinoff from a spinoff. Suntiger has now advanced the technology to embrace a new line of third generation spinoff applications.

Among them are industrial inspection glasses designed to protect plant workers from impact and splash hazards as well as harmful levels of ultraviolet and blue light. This type of glass is especially useful, Suntiger says, for inspection of parts by fluorescent dye testing, which emits hazardous radiations. This brings the PST technology full circle, because it evolved from an industrial project intended to develop a protective welding curtain that filtered out harmful irradiance. That work was done by James B. Stephens and the late Dr. Charles G. Miller, both of Jet Propulsion Laboratory (JPL). Working on their own time, applying radiation know-how and problem solving methodology from their JPL experience, they developed a formula that includes light filtering dyes and small particles of zinc oxide, and produced a commercially-sold protective curtain that absorbs, filters and scatters light. That success led to further research, by Stephens and others who had assisted in the curtain project, in the field of protective glasses.

In addition to the new industrial inspection glasses, Suntiger research on the relative hazards of different wavelengths of light has led to development of safety glasses for protection against certain types of laser light; the company has been able to achieve very high optical densities at the wavelengths emitted by certain lasers without sacrificing visibility.

Suntiger has designed one type of protective lens for dentists (right) who use ultraviolet curing systems, which typically generate hazardous levels of blue as well as ultraviolet light. The lenses attenuate the radiation to safe levels while maximizing the amount of non-hazardous light so that the dentist can see more clearly.

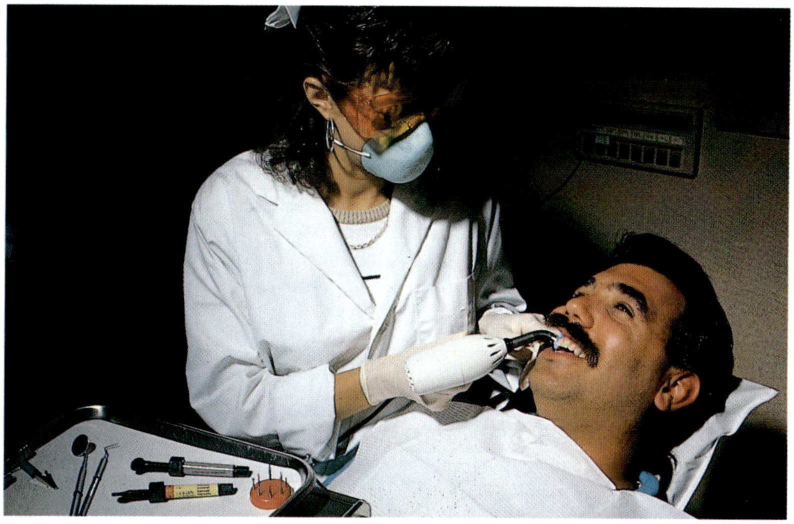

Still another new product is the Fluorotech™ lens (right) specifically developed to block the hazardous radiation peaks emitted by fluorescent lights and unfiltered CRT screens, which have caused complaints of eye irritations and headaches (see page 68).

In development is a line of ski visors that will incorporate polarized orange technology along with several features designed to minimize lens fogging. The original line of PST lenses is being offered in several colorations with four newly-developed frame styles. Suntiger also coats prescription lenses.

® Suntiger is a registered trademark of Suntiger Inc.

™ PST and Fluorotech are trademarks of Suntiger Inc.

Cleaner Air for Home and Office

If often happens that an improvement in one direction spawns difficulty in another. Take, for instance, the U.S. trend of recent years toward superinsulated homes and offices, which has undoubtedly increased energy efficiency substantially but has heightened the problem of indoor air pollution.

The problem is that sealed buildings have less exchange of fresh outdoor air for stale indoor air. This causes higher concentrations of toxic chemicals in indoor environments, brought about by emissions from a great variety of building constituents such as synthetic wallboard, synthetic fibers and glues, cleaning products and insecticides, and gas or woodburning appliances. The result is a notable increase in allergic reactions and other toxicity-occasioned illnesses.

Natural air purification systems based on space technology lead a sampling of environment related spinoffs

Space technology offers an answer to this problem, a solution based on Nature's process of photosynthesis in plant life. Plants generally "breathe in" carbon dioxide and give off oxygen and water. But recent NASA research shows that certain plants can absorb other gases from the air and thereby reduce indoor air pollution. Further, NASA studies show that combining plant foliage with a bed of activated carbon creates a filtration system with a cleansing capability significantly greater than plants alone. This opens the door for commercial marketing of home/office filtration systems.

Why did NASA get into plant studies?

Because plants offer a possible solution to a problem associated with future interplanetary manned spacecraft, space stations or space colonies. The occupants of such spacecraft are protected from the airlessness and extreme temperatures of space by an airtight artificial atmosphere in a sealed vessel. But the spacecraft could not carry enough oxygen and water for normal supply of a large crew for months or even years. Therefore, the initial supply of oxygen and water must be cleansed, detoxified and used over and over.

For more than 15 years, Dr. Billy C. Wolverton and his aides in the Environmental Research Laboratory of John C. Stennis Space Center (SSC), Mississippi, have been conducting innovative research employing natural biological processes for air and water purification. The long range goal is development of a bioregenerative life support system for long duration spacecraft. But Wolverton and his SSC team are also playing an important part in NASA's Technology Utilization Program, which seeks to expand spinoff applications of NASA-developed technology; they have been notably successful in that regard.

In 1974, Wolverton began exploring "aquaculture," the use of aquatic plants to remove pollutants from wastewater at relatively low cost. Initial focus was on use of the water hyacinth which literally thrives on sewage and can absorb astonishing amounts of pollutants. After successful tests at SSC, the facility's neighboring community of Bay St. Louis became — in 1975 — the nation's first municipality to employ an operational aquaculture filtration system. Since then, a number of U.S. towns have adopted hyacinth-based aquaculture as their year-round primary method of treating wastewater. Other towns — and one major city, San Diego — use aquaculture as a supplementary process in sewage treatment.

While the water hyacinth is highly effective means of purifying wastewater on Earth, its utility in space applications is limited. So Wolverton's group developed a more effective technique for in-space water reclamation and toxic chemical removal: the artificial marsh filtering system, which employs a combination of sewage-digesting microbes living in a rock bed and pollutant-absorbing plants such as bulrushes, reeds and canna lilies.

This second generation aquaculture system offers a bonus in Earth applications: where water hyacinths are warm climate plants, which restricts their use to southern U.S., the types of plants used in the artificial marsh system are cold- and salt-tolerant, thus capable of being used in wastewater systems in colder climates. The artificial marsh has already been adopted by some communities and it seems likely that this new technique will spread as widely as its predecessor.

Branching off from aquaculture for wastewater treatment, SSC began exploring the use of foliage plants for air filtration and purification in both space and Earth applications. Wolverton's group evaluated the ability of certain plants to remove the three most common pollutants in tightly insulated buildings: formaldehyde, benzene and carbon monoxide. They found that philodendrons, golden pothos, the common spider plant, Chinese evergreens and others are particularly effective. But this work is still in progress and Wolverton states that any conclusions are premature until all the plants are tested against a wide range of pollutants.

However, another NASA study indicates that a carbon/plant filter system — wherein a bed of activated carbon helps plant roots absorb pollutants — can remove high levels of toxic chemicals and tobacco smoke. Entrepreneurs are borrowing this NASA technology and at least two companies are offering such filtering systems in the commercial marketplace.

Exterior and interior views of the Bio-Home at Stennis Space Center, where NASA researchers are exploring the capabilities of certain plants to absorb gases and reduce pollution in long duration spacecraft or in superinsulated homes and offices.

(Continued)

Cleaner Air for Home and Office

(Continued)

A research assistant is conducting an absorption test; the plant is placed in a sealed chamber into which a gas is injected, then the amount of gas absorbed by the plant is measured. Stennis' Environment Research Laboratory is testing a variety of plants against a wide range ofindoor pollutants.

"Energy-efficient structures are often 10 times more polluted than the air outside," says sales literature from Bio-Safe Inc., New Braunfels, Texas, manufacturer of hybrid plant-microbe filters based on technology developed by the NASA Environmental Research Laboratory at Stennis Space Center.

The reason is that a score or more of interior-use products, from insecticides to shoe polish, plywood board to paper towels, fire retardant materials to permanent press clothing, exude a variety of synthetic chemicals that might not be found in the outside atmosphere.

Bio-Safe's products go a step beyond the growing practice of "interiorscaping" homes and office buildings, hotel lobbies, restaurants, hospitals and other structures, improving the visual aspect by use of decorative live plants. Bio-Safe's systems are designed to remove from indoor air the principal toxic substance formaldehyde, which is present in virtually all modern living environments; tobacco smoke (carbon dioxide) that cannot be entirely removed by plants alone; and large quantities of other pollutants.

The Bio-Safe design consists of a stone white matt finished pot; a plant, such as the split leaf philodendron; a bed of activated carbon (charcoal); and an air pump, or fan, installed near the root system of the plant.

The pump draws room air into the plant-microbe system, pulling it through the charcoal and over the roots of the plants. The pollutants are trapped by the charcoal and digested by the plant's roots or broken down by microorganisms living in the roots. The pump then directs purified air back into the room.

Applied Indoor Resource Company, Tampa, Florida, also used the NASA technology as a departure point for a bioregenerative air purifier but came up with a slightly different approach.

Marketed under the label Bio-Pure, the system includes a foliage filter plant in a planter and, beneath the plant, a layer of patented soil medium — called Dandy Dirt — with activated carbon, legumes and mosses serving as air filtering agents. A mechanical blower moves air through the filtering system for cleansing by microorganisms. Bio-Pure effectively deals with formaldehyde, smoke, food and animal odors and is being tested for radon removal capability. Bio-Pure portable purifiers come in four planter sizes from eight to 20 inches.

Both Bio-Safe and Applied Indoor Resource Company began marketing their filters in 1988. Both reported good initial public reaction.

Shown in his workshop is Jack Reber, president of Bio-Safe Incorporated, who is using NASA natural filtration technology to produce plant systems that remove hazardous pollutants from indoor air. Modern building materials toxic chemical and these pollutants build up in superinsulated sealed environments.

Dr. Wolverton and his group at Stennis Space Center are continuing their exploration of natural air purification systems, testing a variety of plants in a windowless, highly-insulated quonset-like facility. The Associated Landscape Contractors of America are participating in the plant research and the results will be passed along to the association's member companies.

In time, the bioregenerative air purification system may be applied to cleansing whole office buildings. Wolverton's group has explored the possibility of scaling up the system and drawn up some concepts of "atmospheric revitalization" by natural purification. Example: on the roof of an office building is a large penthouse green garden. Stale air from the offices below is drawn into the garden through an intake, channeled through the garden for cleansing, then returned through a series of ducts to the offices. Such a system would enable architects to design buildings for high energy efficiency without worrying about the toxicity effects of superinsulation.

Mrs. Doug Reber is assembling a Bio-Safe Plant system, topping a layer of dirt with a layer of charoal that helps the plant—next to go into the container—absorb contaminants. The system includes an air pump that pulls in room air, routes it through the charcoal and the plant's roots, where the pollutants are trapped and idgested; the pump then sends the cleaned air back to the room. Below, a typical Bio-Safe system.

Here Jack Reber is using a special instrument to check the level of contamination in a client's office.

Gas Analyzer

At right center, a technician in protective gear is using a M200 Microsensor Gas Analyzer to check out a suspected hazardous container. Developed by Microsensor Technology Inc. (MTI), Fremont, California, and derived from space technology, the M200 features dual gas chromatographs, systems that separate a gaseous mixture into its components and measure the concentrations of each gas in the mixture. Such systems are widely used by research organizations and in such industrial applications as monitoring work areas for gas leaks or volatile chemical spills; monitoring stack gases for compliance with pollution laws; identifying gases produced during energy explorations; in police work, for breath alcohol analysis and arson investigations; and in medicine, for respiratory and anesthiology analysis.

The M200 Microsensor is a second generation spinoff with a number of advancements over its predecessor, in particular the dual chromatograph feature. Innovative micromachining technology enabled MTI to fabricate a gas chromatograph on a silicon wafer and overcome design limitations that had earlier prevented production of a high speed, high resolution chromatograph. The M200 completes many analyses in less than 30 seconds; it has the ability to analyze a very large range of mixtures and to measure concentrations as small as one part in a million. Intended for use in plant, field or lab oratory, the M200 weighs only 12 pounds. At right are the detector unit and the gas chromatograph column of the miniaturized M200.

The M200 traces its origin to work performed in the early 1970s by Ames Research Center, which sought to develop a gas analyzer for two unmanned Mars-exploring spacecraft known as Viking Landers. Ames wanted a highly sophisticated gas chromatograph for detection of life forms, if they existed on Mars, and for analysis of the Martian soil and atmosphere. But space was at a premium in the Landers and therefore the chromatograph had to be extremely small and lightweight despite the exceptional performance demanded of it. Ames contracted with Stanford University for hardware development of the instrument.

The unit was not developed in time for use aboard the Viking Landers, but the technology interested the National Institute for Occupational Safety and Health (NIOSH). Looking for a portable device capable of detecting gas leaks in industrial environments, NIOSH provided funds for further development of the Ames/Stanford chromatograph. Subsequently, three researchers who had worked on the project left Stanford and formed MTI to produce a portable gas analyzer for the commercial market. They introduced the orginal version, called Michromonitor 500, in 1982 and the more advanced M200 in 1988.

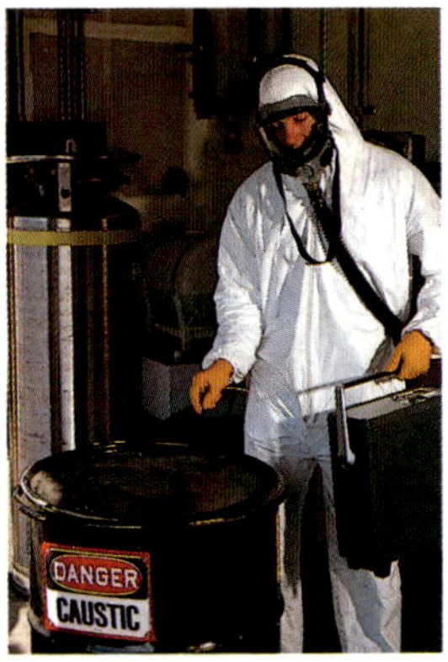

Pollution Control Device

In the accompanying photo, the unit at left is a PrecipTech Power Guard SQ-200 Automatic Voltage Controller which, combined with a remote computer, represents an automatic management system for smoke-cleaning precipitators employed for air pollution control by smokestack industries. The Power Guard SQ-200 is an offshoot of an original SQ-100 developed in a research program conducted by Langley Research Center. The latter system was jointly designed and developed by Kinetic Controls, Inc., Newport News, Virginia and PrecipTech Inc., ESP Specialties, Kansas City, Missouri; PrecipTech now markets the control.

An electrostatic precipitator cleans smoke by removing particulate matter from smokestack gases before the gases are expelled into the atmosphere. Smokestack gas is passed through a precipitator chamber and exposed to an electrostatic field; dust particles in the gas become electrically charged and migrate to collection surfaces where they are "captured." For maximum particle collection, the precipitator must be operated at the highest possible electrical field strength.

Sparking, a normal function of electrostatic precipitation, can limit collection efficiency if not controlled. The Power Guard SQ-200 automatically monitors sparking and quenches sparks before arcing develops; it also automatically adjusts to changes in operating conditions. The unit's ability to fine-tune itself after initial set-up virtually eliminates the need for monitoring by highly skilled operators. It has demonstrated reliability on precipitators in cement plants, paper mills, steel mills, utilities and refuse incinerators.

The initial precipitator control was invented in the course of a 1980 program seeking a way to dispose of refuse in areas where acceptable landfill sites are scarce. Langley Research Center, Langley Air Force Base and the adjoining City of Hampton, Virginia teamed in a highly successful effort to develop a Refuse-fired Steam Generating Facility that incinerates trash, reduces it to a readily disposable ash, and recovers the heat of trash-burning to create steam for practical use at Langley Research Center.

The control evolved as an answer to a problem of excessive stack emissions. When refuse is used as a fuel, the chemical composition of the exhaust gas in the smokestack changes continually — and that requires a voltage controller capable of following process changes. It was necessary, therefore, to develop a control system that matched voltage to smoke changes to insure minimal air pollution.

Two NASA employees — David F. Johnston and T.K. Lusby, Jr. — undertook to develop the control, working for the most part on personal time and with private funds. They successfully produced an innovative, microprocessor based control that automatically senses and compensates for process changes by adjusting precipitator voltage and current for maximum particle collection. In 1987, Johnston formed Kinetic Controls to commercialize the invention.

Wind Generators

At right, hilltop ridges near San Francisco are covered with wind generators, or "windmills," of various designs. This "wind farm" is typical of facilities around the world that utilize wind energy to drive a turbine that generates electricity for use by utility companies. Many of these wind systems employ an advanced generator controller—manufactured by Enerpro Inc., Goleta, California—that incorporates technology originally developed at Marshall Space Flight Center (MSFC) as part of NASA's energy conservation research of the 1970s.

MSFC engineer Frank Nola invented a way of curbing power wastage: a device called the Power Factor Controller (PFC) that matches voltage with an AC motor's actual need rather than its standard fixed voltage, which is needed to handle the heaviest loads the motor is designed to carry. The PFC continuously determines motor load by sensing shifts in the relationship between voltage and current flow; when it senses a light load, it cuts the voltage to the minimum needed, which also reduces current flow and heat loss. With great potential for energy savings over a very broad range of applications, the PFC has become one of the most widely adopted technology transfers.

In 1985, Enerpro president Frank J. Bourbeau (right) conceived a novel means of synchronizing a wind generator to the electric utility grid. In testing the concept on a wind farm machine, he found the generator's controller to be unstable. He devised a way of stabilizing the controller and filed for a patent, but in the course of the patent search he learned of the basic operating principle invented by MSFC's Frank Nola for use with single phase generators. Bourbeau advanced the basic technology by refining the controller design to permit its operation with three phase generators. Bourbeau subsequently received—in 1987—a NASA license for the MSFC technology and was awarded a U.S. patent for his improvements to the Nola invention. The Enerpro device, known as the Auto Synchronous Controller (ASC), is shown below. Enerpro builds all components of the controller in its own facility; at right a company technician is checking a control circuit board and at right below Enerpro engineers are computer testing an ASC.

A primary design objective of the ASC is to reduce generator "inrush" current, a short duration of typically 12 to 15 times rated current that occurs when large generators are brought on line abruptly. The ASC controls the voltage applied to the generator so that the generator is smoothly connected to the utility grid when the generator reaches its synchronous speed. This protects generator components from damage caused by inrush current.

The ASC also increases generator efficency in light winds. Wind powered generators are sized to produce rated power at wind velocities of about 30 miles per hour; as a result, they are considerably oversized for the more typical wind velocities—10 to 25 miles per hour—usually encountered in wind farm environments, hence they are inefficient when operating in light winds.

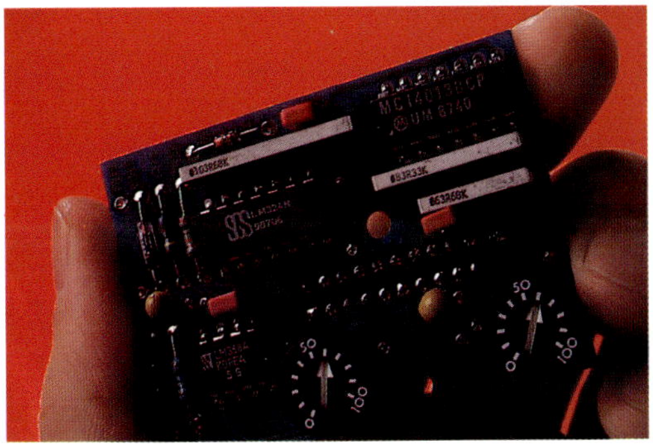

The ASC applies lower than rated voltage to the generator in light winds; this in effect converts the oversized generator into an electrically "smaller" machine, reducing internal electrical losses and increasing power output.

The Enerpro ASC won a 1987 award for energy conservation from the California Energy Commission. Enerpro's first sale—in 1985—was to Carter Wind Systems, Burkburnett, Texas and since then the company has delivered some 1,500 units, including installations in The Netherlands, West Germany, the United Kingdom, Sardinia and Kenya. Enerpro recently won a major contract from Cannon Energy Corporation for 36 controllers to be installed on 200 kilowatt wind turbines at a site near Tehachapi, California.

Space Technology for Patient Monitoring

Headquartered in Clarence, New York, Mennen Medical Inc. is one of the world's leading manufacturers of patient monitoring equipment for hospitals. The firm has a main plant in Clarence, another in Israel, and sales offices in North America, Europe and Australia. Its systems are installed in medical institutions all over the world, including China and the Soviet Union. The company employs some 500 people and records sales on the order of $40 million annually.

Among spinoffs in the field of medicine is a steadily-growing company producing a variety of medical electronics for improved patient care

All that started with a contract for development of an astronaut monitoring system in the early days of the space program. That work provided a foundation in telemetry and other physiological monitoring technologies that led to development of a broad line of computerized medical electronic systems used by hospitals in intensive care units, operating and recovery rooms, neonatal and pediatric units, emergency rooms, shock/trauma units and special procedure facilities.

The guiding hand behind Mennen Medical's consistent growth is that of founder, president and chairman of the board Herbert Mennen. An electrical engineer, Mennen started his career in 1950 with Bell Aircraft Corporation, Buffalo, New York, where he helped develop the nation's first fully automatic aircraft landing system.

In 1957, Mennen and three other Bell engineers formed Sierra Research Corporation and Mennen served as its vice president and director of engineering for several years. It was during this period that he met Wilson Greatbatch, who had invented a cardiac pacemaker, and the two teamed on the astronaut monitoring system. The technologies they developed in that program had clear applicability to another important need — round-the-clock monitoring of hospital patients from central stations equipped with displays of vital signs transmitted wirelessly from bedsides, reducing the time needed for personal attendance by physician or nurse.

"We decided that the real application was on Earth and not in space," says Mennen. "There are only so many astronauts but there are a lot of hospitals."

That decision prompted formation by Mennen and Greatbatch

— in 1963 — of a company that continued their work on space/ medical instrumentation but also initiated a pioneering effort in medical electronics. Greatbatch subsequently left the firm to pursue other interests.

Mennen Medical was the first to adopt total solid state design in patient monitoring equipment and the first to offer multipatient telemetry monitoring. In 1971, the company introduced the first computer-assisted monitor for multipatient arrhythmia, or erratic heart action. More recently — in 1982 — Mennen Medical scored another "first" when it concluded a multimillion dollar sale of equipment to several Soviet hospitals, marking the first use of American-built patient monitoring systems in the U.S.S.R.

A Mennen software engineer compares notes with a hardware engineer.

The key technology in the Mennen Medical line is the space-developed art of telemetry, in which instrument data is converted to electrical signals and relayed to a remote receiver where the signals are reconverted to display information. In patient monitoring, for example, heart readings acquired by an electrode are sent by wire to a telemetry transmitter attached to the patient's body, then relayed wirelessly to a display console at a central station, where a nurse can simutaneously monitor the conditions of several patients.

Herbert Mennen (left center) with some of the staff of Mennen Medical Inc., a spinoff company that traces its origins to a research contract for an astronaut monitoring system.

A typical Mennen Medical product is the telemetry-based VISTA Nurse Station, which presents cardiac waveforms and other clinical information on high resolution color video displays. Designed to provide a full range of patient information that is easy to read and easy to interpret, VISTA systems employ advanced display technology featuring color presentations; the colors indicate the relative urgency of the various monitoring messages, with bright red alerting the hospital staff to alarm conditions. VISTA Nurse Stations come in four, six and eight-patient versions, which can be combined to accommodate any number of patients.

Other examples of company products include the Horizon 2000 Monitor, featuring high capacity computation, analysis, storage and display power; the Horizon 2110 Neonatal Monitor for clinical monitoring of newborns in intensive care units; and the Horizon 9000, a highly sophisticated, fast acting computing system for the cardiac catheterization laboratory.

Mennen Medical's future looks bright to Herbert Mennen. The firm has been growing steadily in recent years, with sales growth of 10–15 percent annually, and, says Mennen, "I see no reason why that pattern won't continue."

A quality assurance technician performs a final inspection on a Horizon patient monitor.

Invisible Braces

The young woman in the lower photo is wearing new type dental braces that are virtually invisible at normal contact distances; only in closeup, as in the upper photo, are the translucent braces detectable.

They are Transcend® ceramic braces, jointly developed by Ceradyne, Inc., Costa Mesa, California and Unitek® Corporation/3M, Monrovia, California, with an assist from the NASA Industrial Application Center, University of Southern California (NIAC/USC), Los Angeles California. NIAC/USC is one of 10 NASA-sponsored dissemination centers that provide information search and retrieval services to industry clients and offer assistance in applying the information. Ceradyne is a regular user of NIAC/USC services.

Marketed by Unitek/3M, Transcend Brackets® represent a high technology orthodontic innovation in which individual translucent brackets, especially designed for each tooth, work in concert with a thin metal connecting wire to gradually reposition teeth, mouth and jaws into proper alignment. Intended to meet a need for an orthodontic appliance that was aesthetically appealing yet as clinically effective as plastic or metal braces, Transcend brackets are made of a very hard, shatter-resistant alumina with high strength and maximum translucency. The translucency allows light to pass through the ceramic material to the tooth, thereby causing the bracket to appear tooth-colored. The brackets do not stain, discolor, deform or bend.

The material is known as translucent polycrystalline alumina, or TPA. It came to orthodontics by an indirect route. Ceradyne, a leader in advanced ceramics for defense, aerospace, electronics and industrial uses, was looking for a special material to be used in infrared radomes employed by the military services in tracking heat-seeking missles. TPA emerged as a leading candidate. At Ceradyne's request, NIAC/USC conducted an extensive literature and patent search to provide a technology base for Ceradyne production of TPA.

In 1986, Unitek contacted Ceradyne in quest of a transparent material of sufficient tensile strength to be used in orthodontic treatment. Ceradyne suggested TPA as the answer and the two companies embarked on a program of development and clinical trials. Transcend Brackets were introduced in 1987 and in the same year production soared to 300,000 pieces a month, marking what its developers say was the most successful orthodontic product introduction in history.

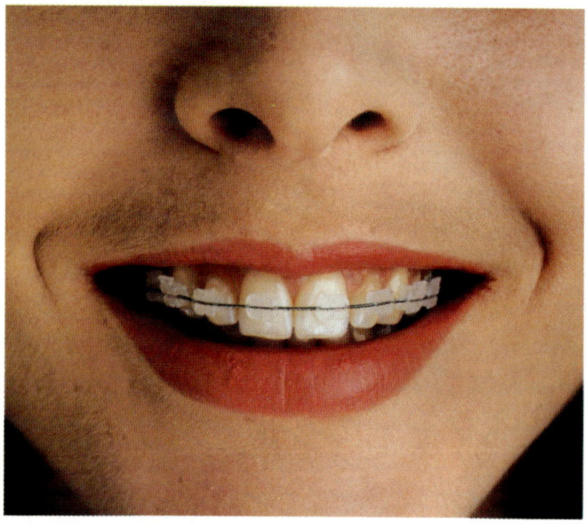

® Transcend Brackets and Unitek are registered trademarks of Unitek Corporation/3M.

Drug Research

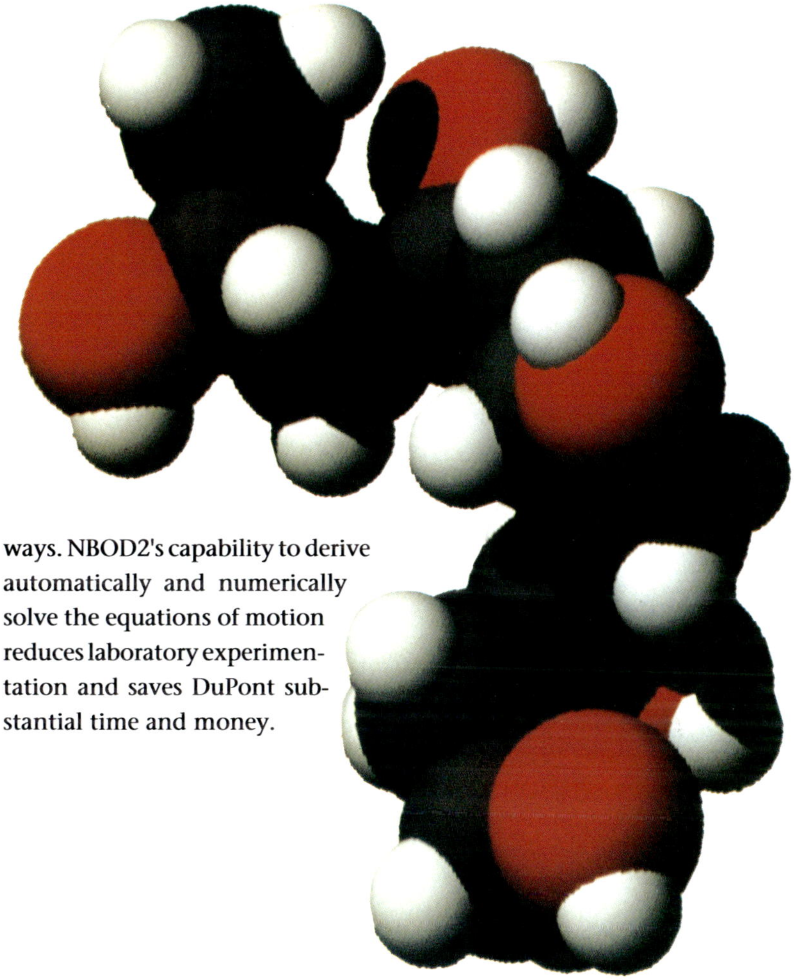

Analysis of the dynamic characteristics of a complex system — a spacecraft, for example — is usually best accomplished by study of a simulation model with the aid of a specially developed computer program. An example is a program developed by Goddard Space Flight Center to solve equations of motion for coupled N-body systems.

Drugs are complex systems, too, and at E.I. DuPont de Nemours & Company, Wilmington, Delaware, researchers are using the Goddard program to model the energy and motion of drug/enzyme bonding as part of the process of designing new drugs. The program was supplied to DuPont by NASA's Computer Software Management and Information Center (COSMIC)®. Located at the University of Georgia, COSMIC routinely makes available to industrial and other clients computer programs that have secondary availability.

Drugs affect biological systems — such as the human body — by interacting with enzymes, complex proteins that catalyze specific biochemical reactions at body temperature. A drug "fits" into an enzyme's receptor slot like a key into a lock and dictates new behavior to the enzyme, thereby affecting the biochemical process that the enzyme catalyzes.

The shape of a drug molecule determines the coupling between the drug and a particular enzyme. Once attached, the drug may change shape and slot into new bonding situations with the enzyme. Each bonding stage has specific biochemical results.

At DuPont, researchers model potential drugs as a series of aggregates (elements) and springs (bonds), as in the accompanying photo. The NBOD2 program then analyzes the vibrational and static motions of these independent components. The information supplied by NBOD2 is used to design specific drugs to interact with particular enzymes in designated ways. NBOD2's capability to derive automatically and numerically solve the equations of motion reduces laboratory experimentation and saves DuPont substantial time and money.

® COSMIC is a registered trademark of the National Aeronautics and Space Administration.

Aid for the Visually Impaired

Sandra Raven is legally blind. She suffers from Stargardts disease, a condition that involves gradual degeneration of the retina and is not correctible although, as in Sandra's case, it may become stabilized. Sandra has lost 98 percent of her central vision but she can see some things with peripheral vision.

Despite her vision impairment, "Sandy" Raven can read, type and hold down a job as clerk typist with the U.S. Coast Guard Training Center, Yorktown, Virginia. The secret is a video system that magnifies and focuses words so that partially sighted people can read and even type from printed copy or handwritten notes. Called the Viewstar, it was invented by Sandy's stepfather, Dr. Leonard Weinstein. "A gift of love," she calls the machine. Above, Sandra is setting up her Viewstar while Weinstein looks on. At right, one of Weinstein's clients, a retired vision-impaired shipbuilder, studies biblical scripture with the aid of his specially-tailored Viewstar; he writes his interpretations with the aid of his Viewstar.

The Viewstar is an example of a personnel-type spinoff, wherein an aerospace engineer or scientist transfers to non-aerospace applications technologies and skills he acquired in his aerospace occupation. Leonard Weinstein is a NASA research engineer, a group leader in the Fluid Mechanics Division, winner of more than 20 awards for innovative research. With degrees in physics, aerospace engineering, fluid mechanics and thermal science, Dr. Weinstein is a man with an exceptional range of expertise; in developing the Viewstar, he combined a number of electronic, optical and mechanical technologies.

Sandra Raven did not lose her sight until

she graduated from Warwick High School, Newport News, Virginia. When her vision deteriorated, she sought help from the Virginia Department of the Visually Handicapped, which loaned her a closed circuit TV visual aid system for magnification. Weinstein, however, was not impressed with the system's performance and it was very expensive; he felt that he could employ his scientific/technological expertise to develop a system of at least comparable performance that would be considerably less costly.

He succeeded. In 1982, he produced the first version of a dual field Viewstar, which offered lower cost principally because it employed only one TV camera where commercially available systems used two. The camera focused on the copy to be typed and presented it in magnified form on the upper half of a split screen monitor; material on the typewriter was shown on the lower half, similarly magnified. The two views could be independently magnified and focused, and a movable mirror allowed use of the system with a moving ball typewriter. Weinstein also produced alternative systems for single view reading, writing or examining small objects.

The original system was about one-third cheaper than available commercial counterparts but Weinstein undertook a second phase development to produce a Viewstar Economy Model, a low vision closed circuit TV system that includes a TV camera, zoom lens, a stand with a lamp for viewing and a 12-inch monitor. It sells for $1,150.

Weinstein also developed other systems to meet individual needs — different magnification ranges, positioning stages, black and white image reversal and larger monitors.

The products are marketed by Viewstar, Newport News, Virginia, a small business owned by Weinstein's wife; he serves as technical director. The company has sold systems to several U.S. universities, state and federal government agencies, private corporations and individuals.

Production is limited, however, because Weinstein still works full time at Langley Research Center and can devote only 10–12 hours a week to producing Viewstars and other specialty items he has invented. "I'm not interested in building a big business," he says, "I just want to help the visually impaired."

Physical Therapy Machine

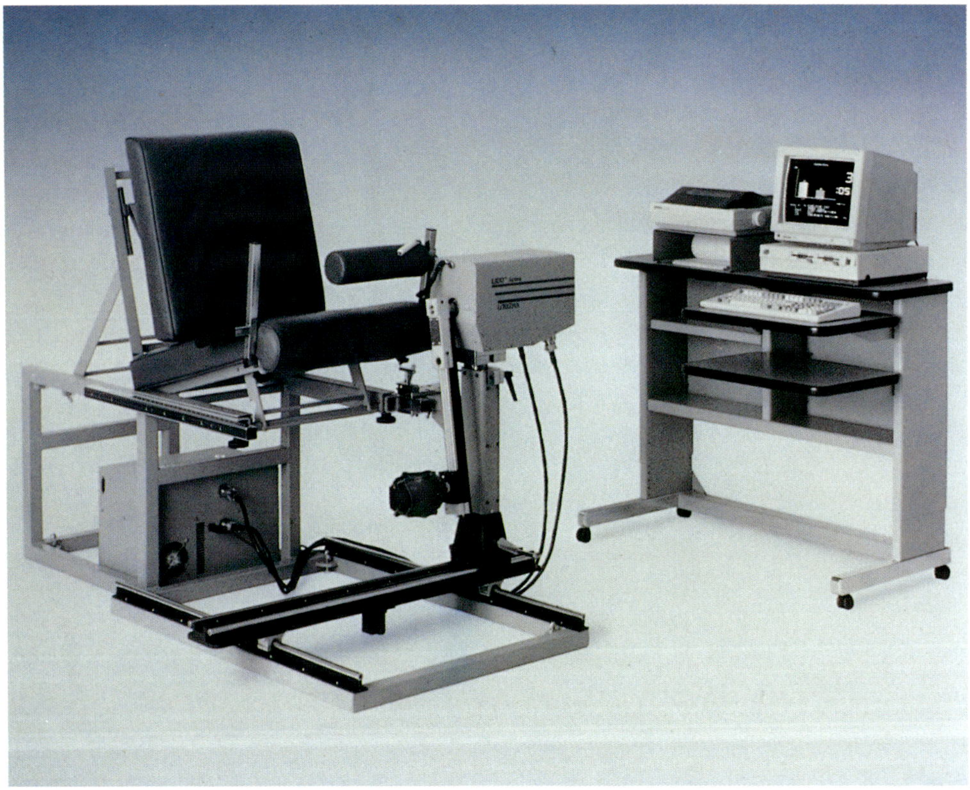

At right are the LIDO® Active (left unit) and the LIDO Digital computerized physical therapy machines, used by clinics, medical research centers, sports medicine facilities and sports teams for evaluation and rehabilitation of wrist, elbow, shoulder, hip, knee and ankle musculature.

The machines are produced by Loredan Biomedical, Inc., Davis, California. The LIDO Digital system is employed in isokinetic concentric exercise programs. The LIDO Active builds on the capabilities of the LIDO Digital and offers four additional exercise modes. Both Loredan's systems use LIDOSOFT highly interactive software package, the spinoff element of the LIDO systems.

The systems were developed by Malcolm L. Bond, Loredan chairman and chief executive officer, who holds a doctorate in physiology from the University of California-Davis and who conducted post-doctorate research at the University of Padua (Italy) Center of Muscle Pathology. Bond spent a decade studying muscle and tissue diseases before he developed the technology for his LIDO systems. Loredan designs the systems, farms out the manufacturing work on components, and assembles the systems at the Davis facility.

In 1985, as part of its preliminary planning for the Space Station *Freedom*, NASA bought a LIDO system, along with a number of muscle therapy devices designed by other manufacturers, for evaluation as a Space Station aid. Living extended periods under the weightless conditions of orbit, astronauts could suffer muscle and tissue deterioration unless they conduct regular exercise programs with the help of specially designed physical therapy equipment.

In 1986, Ames Research Center conducted a muscle conditioning project in which a large number of volunteers, bedridden 24 hours a day for 30 days to simulate the effects of weightlessness, were rehabilitated by various exercise programs and physical therapy devices. This project provided an informational base for selection of the exercise programs and equipment to be used aboard the Space Station.

Malcolm Bond's LIDO and Bond himself participated in the program. To meet NASA requirements, Bond designed an advanced software package that became the basis for the

LIDOSOFT software used in the commercially available Loredan systems.

The LIDO systems employ a "proprioceptive" software program, meaning one capable of perceiving internal body conditions, such as the angle and rate of speed at which a patient uses his muscles. The computer program can also induce perturbations to human muscular effort — changes in force, joint angular velocity, joint position — and it can measure and evaluate the response. Real time biofeedback, presented on a screen in easily understood graphics, allows a patient to observe his own performance.

At far right, a patient is using a LIDO Active, which is changing the exercise load, tailoring it to the patient's changing capabilities and fatigue level during an extended exercise "bout." The upper right photo shows Loredan's patented sliding cuff, which helps decrease the compressive forces on the joints and thereby significantly improves patient comfort during an exercise program. At right center, a patient is using a Loredan WorkSET, employed for simulation, evaluation and training/rehabilitation of work-related injuries.

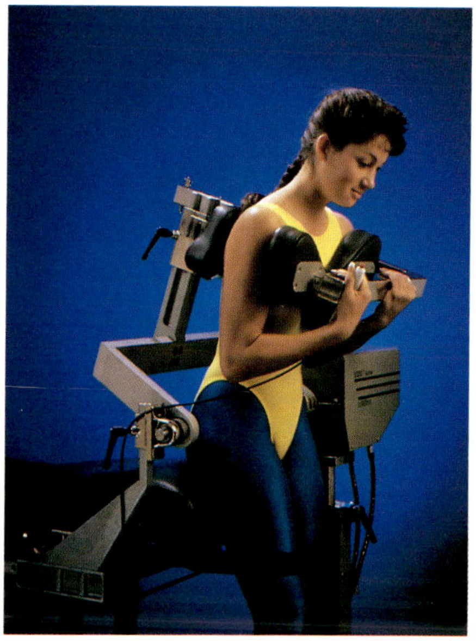

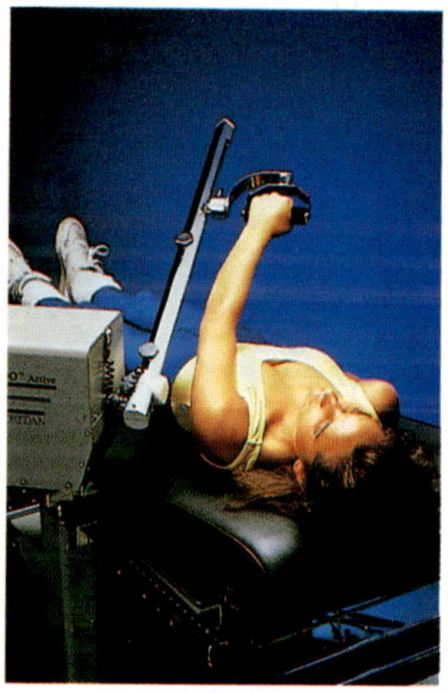

® LIDO is registered trademark of Loredan Biomedical, Inc.

Clean Room Apparel

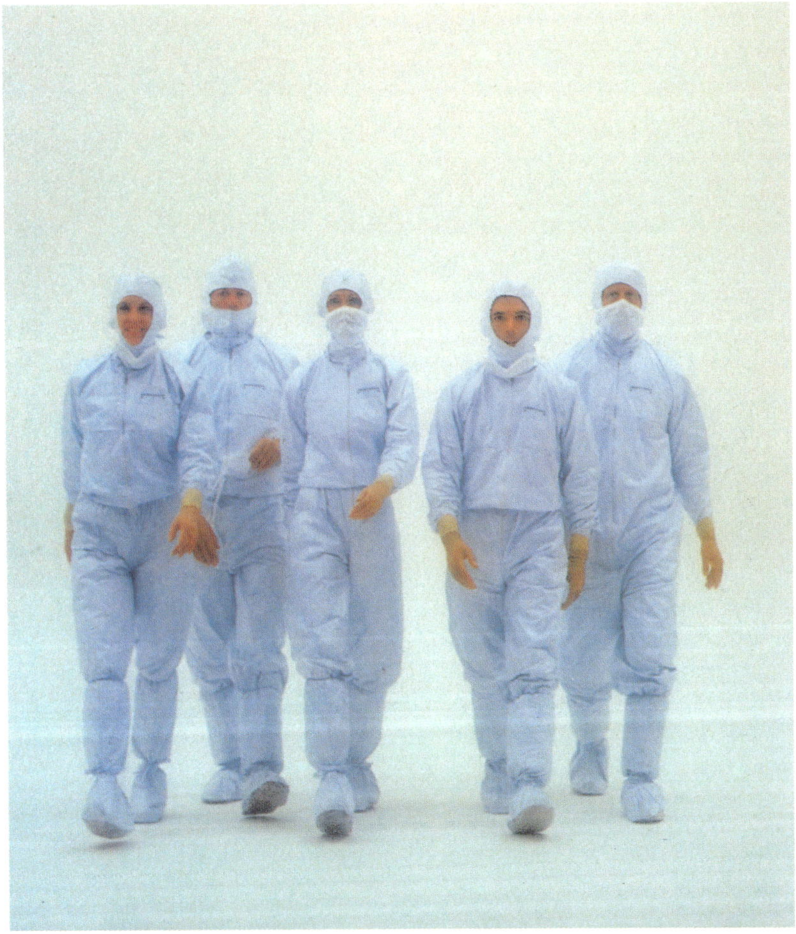

The people at right are displaying a line of contamination control garments used by hospitals, pharmaceutical and medical equipment manufacturers, aerospace and electronic plants, and other industrial facilities where extreme cleanliness is important. They are produced and marketed by Baxter Healthcare Corporation, Industrial Division, Valencia, California under the trade name Micro-Clean® 212.

The lower left photo is a closeup of the "high top" shoe covers that extend all the way to the knee. At lower right is an open face hood and face mask. At far right is the clean room overall with accessories.

The Micro-Clean 212 garments represent a second generation spinoff, an advancement over an original Micro-Clean line introduced in 1982 and based on NASA contamination control technology.

NASA began developing such technology in the earliest days of the space program because delicate instrumentation and sensitive electronic systems are highly susceptible to breakdown through contamination. A tiny mote of dust, introduced to the equipment during its fabrication, can trigger a malfunction that can at best impair a system's accuracy and precision, at worst cause mission failure.

To bar such occurrences, flight equipment is assembled in clean rooms that match or surpass hospital standards of cleanliness. The air entering these facilities is filtered, temperature and humidity precisely controlled; the rooms are designed to eliminate nooks and crannies where dust particles might collect. Workers wear special lint-free clothing and they enter the clean room through an airlock that prevents contamination from outside air.

To help its contractors set up clean rooms and develop advanced control measures, NASA spearheaded contamination control technology, building an informational base with input

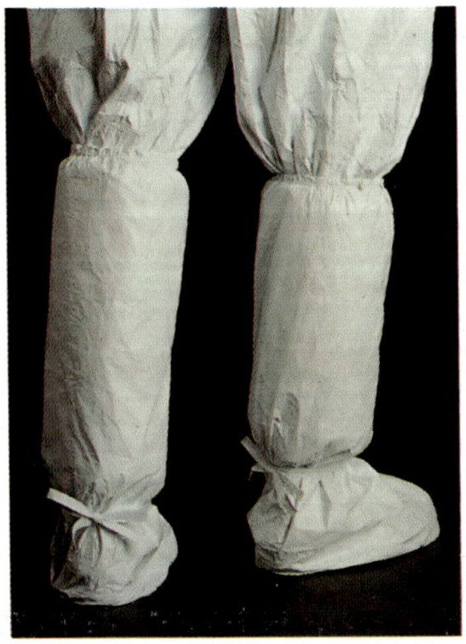

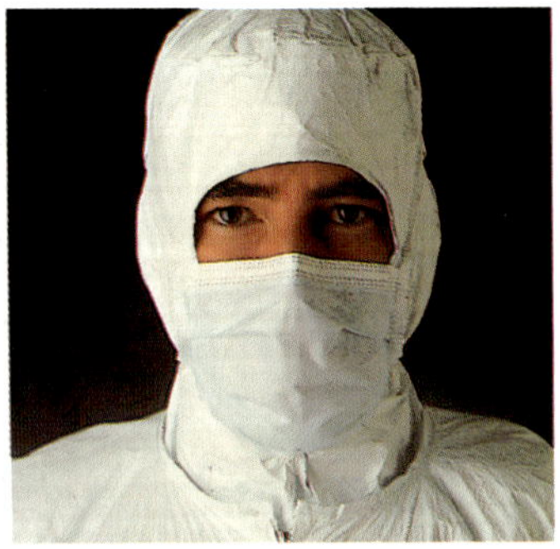

from Marshall Space Flight Center, Johnson Space Center, Kennedy Space Center, Lewis Research Center and Sandia Laboratories. NASA conducted a number of special courses for clean room technicians and supervisors and published a series of handbooks that represented the most comprehensive body of contamination control information available at that time (in the 1960s).

American Hospital Supply Corporation (AHSC), Baxter Healthcare's predecessor company, used the NASA informational base as a departure point for a research project aimed at improving industrial contamination control technology. In 1980, AHSC researchers studied the NASA handbooks, visited NASA centers, and investigated several contractor clean room operations, acquiring a wealth of information on contamination control technology and problem areas.

This research project concluded that the greatest sources of clean room contamination were the people who worked in such facilities; they generated microscopic body particles that escaped through tiny "windows" in the woven garments they wore.

This conclusion led to AHSC's development of the original (1982) Micro-Clean line of apparel, made of non-woven material known as Tyvek™ capable of filtering 99 percent of all particulate matter measuring half a micron (a millionth of a meter) and larger.

Baxter Healthcare has continued to improve the line through advanced technology. The key improvement in the new Micro-Clean 212 line is a proprietary polyimide coating applied to the base fabric (Tyvek) to seal and tie down any loose fibers, thereby minimizing fabric linting and particle generation from abrasion. The coating also provides greater durability. Additionally, the company redesigned its coverall to minimize the stress points along the seams and make the garment virtually tearproof.

Micro-Clean 212 garments are disposable, which eliminates the costs of laundering and repair. They come in sterile and non-sterile forms; sterilization is by gamma radiation. Each garment is individually packaged in a clean room.

®Micro-Clean is a registered trademark of Baxter Healthcare Corporation.

Image Processor

At right, a researcher is operating a Programmable Remapper, a digital image processing machine of novel design and function with important potential for alleviating vision problems encountered by people with retinitis pigmentosa, maculopathy and other vision impairments.

The Remapper is a research tool used to determine how to best utilize the part of a patient's visual field still usable by mapping onto this field of vision with manipulated imagery. If, for example, a person's central vision has deteriorated but peripheral vision is still intact, the image is warped or "remapped" onto the still-functioning portion of the eye. The system's developers plan to advance the technology to create a wearable prosthesis for visually-impaired people.

The Remapper is an offshoot of a NASA program for an image processor capable of simplifying, speeding up and improving the accuracy of pattern recognition in video imagery and to solve thereby problems associated with automated tracking/docking of spacecraft and autonomous planetary landings. The original specifications were drawn up by the Tracking and Communications Division of Johnson Space Center (JSC). The design was accomplished jointly by JSC and Texas Instruments, Inc., Dallas, Texas, working under contract to JSC. Dr. Richard D. Juday is JSC project manager. Dr. Jack M. Younse is Texas Instruments' program manager and Dr. Jeffrey B. Sampsell is the company's manager for development of the Remapper.

Early in the program, the Remapper's potential for application to human low vision problems became apparent and NASA's Technology Utilization Office provided funds to pursue research toward that application. The U.S. Army Missile Command saw another application—a more effective system for "teach-

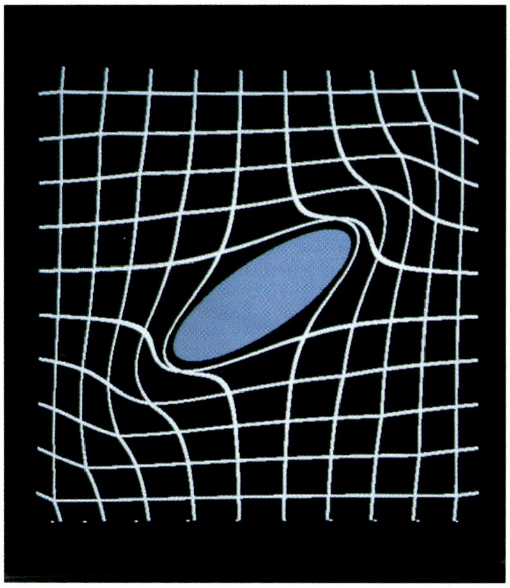

ing" missiles to recognize tanks on the battlefield — and joined the development group. The group is pursuing all three applications — space, battlefield and human low vision — and Texas Instruments is also considering the Remapper as a future industrial robotic vision system.

For the human low vision application, NASA assembled a team composed of Dr. Juday, who is manager of Robotic Vision Development at JSC, and Dr. David S. Loshin of the University of Houston's College of Optometry. In its current form, the Remapper is commercially available from Texas Instruments as a tool for optometric research.

The Remapper makes it possible to "push around" an image so that more of the image structure falls onto the still-functional portions in the retina of a vision-impaired person. Drs. Juday and Loshin are evaluating the potential of remapping as a prosthesis for certain forms of human low vision. At upper left is a grid that has been remapped, meaning that all the information behind the blind (blue oval) area has been remapped outside the blind spot; at lower left, the remapping is demonstrated on a printed page.

The Programmable Remapper works at video rates; what is seen on the monitor screen is what is actually seen by the subject, and no computer analysis is necessary. Thus, the system has strong possibilities as a prosthesis if planned "shrinkage" can sufficiently reduce the size and cost of the device to make it practicable as a portable image correction device. The Remapper, worn on a belt or elsewhere on the body, would warp the image to correspond to the patient's vision characteristics and the viewing task he is attempting; the patient would then view the warped image on a small video display in front of one or both eyes. Drs. Juday and Loshin believe that the essential miniaturization of the system could be accomplished in five years.

The Light That Doesn't Fail

Systems redundancy is a NASA term for the technique of employing one or more backup units in critical spacecraft components, so that if a primary system fails a backup will automatically take over its work. It's a way of assuring the ultrahigh reliability that spacecraft, particularly manned spacecraft, demand.

This space technique has spread to certain Earth-use systems where a failure could be serious and the extra reliability is worth the added cost. But you would hardly expect to find systems redundancy in a home-use flashlight that sells for less than $20.

Flashlight battery innovations typify aerospace technology transfers for consumer, home and recreational applications

Yet that is exactly what Rayovac Corporation, Madison, Wisconsin has incorporated in its Luma 2 flashlight, which company sales literature bills as "The Light Fantastic." Luma 2, a premium flashlight conceived by Rayovac vice chairman Judy D. Pyle, which features an extra-bright Super Krypton primary bulb, has a completely independent backup system that includes a separate lithium power cell, its own bulb and switch. This innovation won top honors among the Wisconsin Governor's New Product Awards in 1988. It is U.S.-patented to inventors David R. Schaller, Sid A. Megahed, James Neyer and Tom Patterson; foreign patents are pending.

Luma 2 is a two-way spinoff. It not only incorporates NASA technology, it also represents an example of the type of innovation-stimulating assistance NASA's Industrial Applications Centers (IACs) provide to industrial clients like Rayovac, a top flashlight and battery designer and manufacturer that supplies standard and customized batteries to retail markets and to manufacturers of tape recorders, radios, headphones, flashlights, hearing aids and other such products.

In addition to its use of the systems redundancy technique, Rayovac also employed in Luma 2 such NASA-developed technologies as the lithium power cell, which offers a 10-year shelf life, and a magnetic switch with corrosion-proof sealed contacts.

The NASA-sponsored IAC that helped Rayovac on Luma 2 and other projects is NERAC, Inc., Tolland, Connecticut, one of 10 IACs that offer industrial clients access to the NASA data bank and some 400 other computerized databases whose combined storage amounts to nearly 100 million documents. The IACs conduct literature searches to find and apply technical information pertinent to the client's needs.

To stay at the leading edge of its technologies and continually improve upon existing products and processes, Rayovac makes frequent use of NERAC's problem-solving and technical support services. With NERAC's help, Rayovac engineers investigate unfamiliar technical areas, identify qualified outside experts in areas of special interest, stay attuned to the latest developments in their disciplines, and glean valuable insights into competitive activity, patents and industry trends.

Shown below is the Rayovac Luma 2 Flashlight, which features an independent backup lighting system in case the primary fails. At bottom is an exploded view showing both bulbs, both batteries and both switches. In addition to employing the space technique of systems redundancy, Luma 2 incorporates NASA magnetic switch and lithium battery technology.

In addition to identifying technology advances incorporated into Luma 2, NERAC also helped the company develop a zinc air battery for hearing aids and improve one of Rayovac's principal battery lines used in such products as tape recorders and radios.

Says Rayovoc senior technical analyst Carrol G. Saxe: "NERAC provides a service to industry that industry could not economically provide for itself. Without NERAC, I estimate that Rayovac would have spent at least $50,000 gathering the information supplied by NERAC — a tremendous benefit."

Scientific Toy

American architecht R. Buckminster Fuller spent a lifetime seeking geometric solutions to architectural problems. Fuller's geodesic dome, an assemblage of triangular shapes, became the basis for a great many homes and commercial structures all over the world.

Fuller's "energetic geometry" was based on fundamental principles of attraction and repulsion found in nature. When tension (pull) is the basis for integrity (completeness), as it is in every molecule, these forces are in "tensegrity," a word Fuller coined. The human body, for example, is a tensegrity structure, kept stable by tensional balances among the skeleton, muscles and connecting tissues.

Fuller's concept and the modeling of tensegrity structures by sculptor Kenneth Snelson inspired the development of an erector-set-like toy, or geodesic puzzle, called Tensegritoy. Designed to give students an understanding of the scientific principles of structural stability, it was created by Stuart Quimby and Cary Kittner, shown above with some of the structures that can be built with the Tensegritoy, including a truss-like display based on space technology.

Quimby and Kittner formed Tensegrity Systems Corporation, Tivoli, New York to market the science toys, used by children aged eight or more and also by architects, engineers and academic science and technology instructors.

A Tensegritoy kit consists basically of a number of wooden dowels or sticks and a roughly equal number of elastic cords. The dowels represent the forces of nature that tend to push matter apart; the elastics, used to connect the dowels, represent the forces that draw matter together. By stretching the elastics from dowel to dowel, it is possible to assemble a great variety of intricate structures that teach a science lesson: the outward force of the dowels would destroy the structure were it not for the

balance created by the inward force of the elastics, and the inward pull of the elastics would collapse the structure except for the outward push of the dowel. The company offers kits in five sizes, ranging from a small introductory kit to one with 300 dowels and associated elastic cords.

Tensegrity Systems also offers mobile and floor stand point-of-purchase displays to retailers handling the Tensegritoy line (right). The floor stand displays are based on technology developed for Space Station trusses, which must be folded into compact packages for delivery to orbit, then deployed by springs into a large truss segment.

Quimby learned of this technology from Tech Briefs, a NASA publication that informs potential users of technology available for transfer (see page 132). Quimby read two Tech Briefs articles describing deployable geodesic trusses that could be collapsed into small packages for Space Shuttle transport, then unfolded in space into a truss 118 times the volume of the compacted package. Quimby used the technology to create mini trusses from Tensegritoy kits that can be deployed into sturdy, lightweight display columns four to 12 feet high. As a result, Tensegrity Systems not only developed an attractive display that takes up only one square foot of floor space, it also realized substantial savings in freight costs. The displays are now shipped in a package only 18 inches high; the display deploys partially on release of an elastic cord and the store owner can complete the assembly in minutes.

At left, Quimby has fully extended a red truss display while a blue truss lies partially deployed in the foreground; to finish the assembly job, he must reconnect stick ends that have been undone for compacting.

Composite Riflescope

At right, an armored "knight" is holding a rifle fitted with an advanced target riflescope made of composite material instead of the aluminum commonly employed in scopes; the imaginative photo is used in advertisements for the Armor-Sight™ riflescope developed by Bushnell Division of Bausch & Lomb.

The Armor-Sight combines Bausch & Lomb's world-renowned optics with a graphite composite — called Graphlon VI — developed for space applications.

"The tube is formed in the same manner as vital space vehicle components," says astronaut Joe Engle, research consultant to Bushnell, in company ads. "Miles of tough graphite fibers woven in a sophisticated matrix are set in polymer resins. The finished product is a riflescope housing that is strong, light, stable and protective of the scope's precise internal optics."

In addition to getting a 10 percent weight reduction in comparison with aluminum scopes, Bushnell got an extra advantage: the material's thermal expansion coefficient is near zero, which obviates problems of optical distortion due to heat expansion and cold contraction. Armor-Sight is fogproof and waterproof, and its advanced multicoated optics deliver maximum light transmission to brighten target images.

The NASA Industrial Applications Center–University of Southern California (NIAC/USC), Los Angeles, California provided an assist to Bushnell in the development of Armor-Sight. NIAC/USC conducted literature searches of the NASA data bank and other data bases to supply the company extensive technical information on graphite composites and, in particular, information of a problem-solving nature that enabled Bushnell R&D engineers to overcome a bonding difficulty and a porosity problem that had cropped up in the course of the development.

™Armor-Sight is trademark of Bausch & Lomb.

Art Preservation

NASA centers conduct extensive research in coating materials for protection of spacecraft and components from the harsh environments in which they must operate. Langley Research Center has been particularly active in this type of work and the center's efforts have produced a number of coatings that have found spinoff applications.

In the mid-1980s, Langley synthesized a new class of polyimides, a group of plastic substances generally noted for resistance to high temperatures, wear and radiation. These polyimides are optically transparent, thermally stable and soluble in some common solvents.

The Getty Conservation Institute's Materials Science Group of the Conservation Research Program, Marina del Rey, California, evaluates new coating and consolidation materials for their possible application to conservation of art objects; at right, Institute scientist Eric Hansen is investigating the properties of a polymer solution. The Institute became interested in the Langley polyimides because of their apparent stability and because they offer deposition of nearly colorless thin films on art objects. The Institutes had two important applications in mind: protection of medieval stained glass windows and prevention of corrosion on outdoor bronze sculptures.

The Materials Science Group evaluated the polyimides, under conditions simulating indoor and outdoor exposures, for changes in color, permeability and flexibility induced by ultraviolet radiation in the atmosphere. Special tests were conducted on stained glass windows in the Cathedral of Leon, Spain and on outdoor bronzes.

The stained glass window application did not prove out but one of the Langley coatings — known as ODPA-3, 3-ODA — appeared after preliminary screening to be a candidate for coating outdoor bronze statuary. The need for protecting bronzes from corrosion involves thousands of pieces from antiquity to recent times in locations all over the world; it is a particular problem in the hot, humid climates of Africa and Asia.

Satellite Imagery via Personal Computer

Tiros 1, the world's first weather satellite launched by NASA in 1960, was a marvel for its day but a technological Neanderthal compared with today's highly sophisticated environmental satellites. Among other limitations, processing the cloud cover pictures was a complicated matter in which the TV camera signals had to be stored and transmitted later to a few Earth stations that had the special equipment necessary to convert the data photographic form.

Highlighting spinoff examples in the field of computer technology is a low cost means of access to weather satellite information

With Tiros 8 in 1963, NASA introduced a major improvement called APT, for Automatic Picture Transmission. APT included an advanced satellite camera that snapped a picture and immediately began transmitting it, plus simplified, low cost receiving equipment for the ground stations. This development made satellite weather images directly and immediately available to anyone willing to make the moderate investment in an APT ground station, thus extending the benefits to weathermen around the world, to commercial TV stations, to colleges and universities, even to private individuals who built their own receivers. By 1966, when the Tiros system graduated from research to operational status, there were more than 300 ground stations, many of them on foreign soil. Eventually, more than 100 foreign nations took advantage of NASA's pledge to make space benefits available "to all mankind."

NASA continued to work on APT and later introduced an advanced scanning radiometer that upgraded the quality of weather pictures but created a problem: most Tiros system users were still operating their original ground station display equipment, which would not readily be adjusted to the new APT format. To accommodate many foreign and other APT users, Goddard Space Flight Center developed an APT Digital Scan Converter that electronically altered the APT data received from a satellite to make it compatible with the older electromechanical display systems.

Goddard's Charles H. Vermillion and John C. Kamoski wrote a comprehensive NASA Technical Note, a voluminous report published in 1975, that described their Digital Scan Converter in complete detail, with construction plans, circuit and wiring diagrams, photos, drawings and dimensional data from which anyone competent in digital electronics could build his own converter with off-the-shelf components readily available in most parts of the world. That Technical Note became the basis for a new product and a complete change of business activity for Electro-Services®, Cleveland, Minnesota, then specializing in maintenance of computer and video systems.

The company used the Goddard technology as a departure point for what became, in 1979, the first microcomputer-based weather imaging system in the U.S. The promise of this system prompted the company to sell off its computer/video maintenance assets in 1984 and embark on a new business mission: design and marketing of a weather image display product for personal computers, including both hardware and software. In 1988, the company's name was changed to Satellite Data Systems Inc. (SDS) to reflect its new activity.

SDS's product is the low cost Electro-Services WeatherFax facsimile display graphics system, which consists of a single ESC-102 plug-in card derived from the NASA technology, software, an instruction manual and an easily-installed connecting cable. The system converts a personal computer into a weather satellite image acquisition and display workstation. A WeatherFax unit for an IBM personal computer costs less than $800.

SDS also offers computer hardware, antennas, receivers and other associated equipment, provides "do-it-yourself" systems and full turnkey systems that include installation and operator training.

The company markets its products worldwide, with installations in the U.S., Canada, Europe and the Far East. A major customer is the U.S. Weather Service; another area of large scale use is educational training in high schools and colleges. Other customers include the U.S. military services, foreign governments, professional meteorologists and amateur hobbyists.

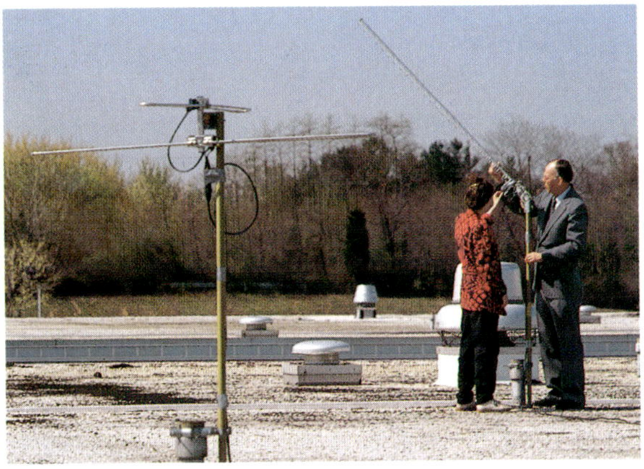

Above, Bob Bowles (right) of Paint Branch High School, Burtonsville, Maryland is instructing his Earth Science class in weather analysis with the help of Satellite Data Services' spinoff WeatherFax a low cost unit that converts a personal computer into a satellite image acquisition and display workstation, thus making satellite signal reception affordable for a wider range of users.
At left, Bowles is adjusting one of the satellite reception antennas on the roof of the school; students and teacher installed the antennas themselves.

Expert System Software

To help boost national productivity, NASA offers assistance to computer-using American businesses by providing a way to effect significant reduction of automation costs: use of government-developed computer programs that can be adapted to secondary usage. NASA's mechanism for making such programs available to the private sector is the Computer Software Management and Information Center (COSMIC)®, located at the University of Georgia.

An example of a widely-used COSMIC program is CLIPS (C Language Integrated Production System), a software shell for developing expert systems. Originally developed by Johnson Space Center, CLIPS is designed to allow research and development of artificial intelligence on conventional computers. CLIPS enables highly efficient pattern matching. A collection of conditions, and the actions to be taken if these conditions are met, is built into a rule network; additional user-supplied facts pertinent to a particular problem are matched to this rule network. CLIPS' versatility has made it a valuable research tool for a variety of applications. Some examples:

Chemical production machines at E.I. DuPont de Nemours & Company, Wilmington, Delaware require constant monitoring of product quality and quantity. The company has embedded a watchdog expert system in the CLIPS shell. Use of CLIPS allows a user to isolate productivity problems and work toward their solutions in the absence of a machine expert.

Under a grant from IBM, the Computer-Aided Productivity Center at California Polytechnic State University is investigating the role of artificial intelligence in computer-aided design. CLIPS has been interfaced to a knowledge base of design rules and solutions. Through question and answer sessions, CLIPS responds to requests for information and provides continuous background monitoring of an evolving design.

Mentor Graphics, San Jose, California used CLIPS as the developmental core of a new Circuit Synthesis System. The system employs a series of "knowledge modules" that provide expertise in disciplines that a user engineer lacks; for example, it allows a *digital* engineer to synthesize many *analog* schematic designs; above, Mentor employee Richard Aikers is using the system to generate a schematic that implements the design specifications he has typed on the screen. The Contour program queries the designer for pertinent circuit data and uses a selected knowledge module to synthesize the schematic. CLIPS acts as an interpreter between the user inputs and the knowledge base.

Tom Brooke of the China Grove, North Carolina law firm of Brooke & Brooke uses CLIPS to help him decide which facts from a casefile are most pertinent and should be included in his legal pleading for the court. CLIPS is interfaced to a word processor and a file of case facts; in question/answer exchange, CLIPS selects facts to be merged with paragraphs in the files that are to be included in the pleading.

® COSMIC is a registered trademark of the National Aeronautics and Space Administration.

Composite Nacelles

The principal products of Rohr Industries, Chula Vista, California are nacelle systems, the aerodynamic structures surrounding an aircraft engine, and pylons, structures that connect the engine to the aircraft either under the wing or projecting from the fuselage. The photo, an exploded view of a nacelle system, shows typical Rohr-manufactured products.

In manufacturing jet engine nacelles, engineers are turning more and more to composite materials, comprised of non-metallic ingredients such as epoxy and graphite fibers, that are lighter yet stronger than the metals they replace. Jamie Abbott, senior thermodynamicist with Rohr, states that his company is near the point of producing an all-composite jet engine nacelle. Rohr is conducting extensive research toward that end.

The most commonly used composites can tolerate heat up to 350 degrees Fahrenheit, but beyond that temperature they lose strength. Since the nacelle embraces a jet engine operating at high temperature, heat flow is a major concern.

To predict heat flow patterns, Rohr acquired a specialized computer program—called TRASYS—from NASA's Computer Software Management and Information Center (COSMIC) at the University of Georgia (see opposite page). Developed at Johnson Space Center, TRASYS aids in predicting how much heat will be generated and how fast it will be dissipated under different conditions.

By using the TRASYS program, Rohr saves on nacelle design costs. The basic nacelle design can easily be modified to adjust for different engines, frames and mounts without extensive testing of heat distribution patterns.

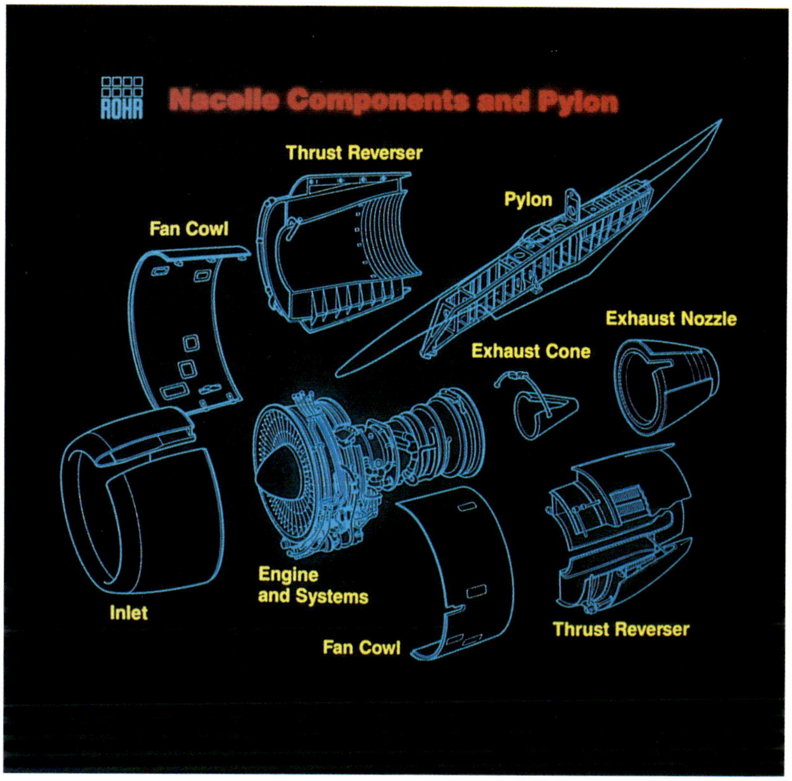

Software Management System

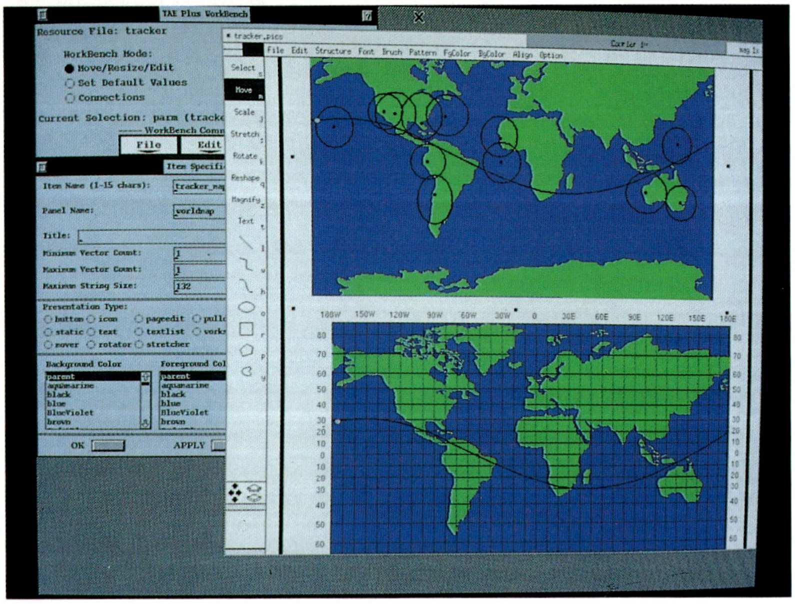

Computer solutions to analysis or information management problems often require use of multiple computer programs. Sometimes the programs are numerous, complex and interrelated. Thus, both system developer and end user can benefit from help in organizing and managing the programs.

A NASA-developed software management system known as the Transportable Applications Environment (TAE) offers such help. It simplifies the job of a system developer by providing user interface development tools and a stable framework in which a system can be built, thereby reducing the time between concept and first implementation. Developed to better serve the needs of end users, applications programmers and system designers, TAE binds a set of applications into a single, easily-operated whole and supports user operation of programs through a consistent, friendly and flexible user interface.

In today's environment, where an application system's software costs usually far outweigh its hardware costs, TAE can provide significant savings. It lowers the cost of system development and software conversion by providing software and structures for commonly recurring user requirements, such as menu and command interfaces, information displays, parameter processing, error reporting and online help.

TAE was originally developed in the early 1980s to support image processing and remote sensing applications at Goddard Space Flight Center. Over the years, it has evolved from a traditional menu and command oriented system to a state-of-the-art user interface development system supporting high resolution graphic workstations (right). The latest version of TAE—called TAE Plus—provides a tool for anyone involved in the building of user interfaces. It is designed for both progammers and non-programmers and there is no coding necessary to build a prototype. TAE Plus allows the application developer to interactively construct the layout of a graphic workstation screen and manipulate a set of "interaction objects," including user entry and information items as well as data-driven graphical objects. For those who intend to develop their prototypes into operational applications, the WorkBench generates source code and resource files (right). With the new TAE Plus user interface design features working in tandem with the original TAE's standard set of application management services, TAE offers an integrated application development tool for increased programmer productivity. At right below, Goddard's Marti Szczur explains TAE Plus to a group from Georgia Tech.

Each year the "transportable" element of TAE becomes more evident as the number of computers on which TAE operates grows. At midyear 1989, TAE was operating on 15 different computers, running under several operating systems. As TAE utilization and functionality grew, so did the range of applications; they now include scientific analysis systems, user assistance teaching tools, defense systems, prototyping; and realtime command and control systems. TAE's versatility had led to widespread use outside NASA; the user community has grown to more than 300 facilities operated by government agencies, universities and industrial firms. A sampling of users includes The Boeing Company; Computer Sciences Corporation; EOSAT, operator of the U.S. Landsat remote sensing system; Harris Corporation; Philip Morris; RAND Corporation; Lawrence Livermore Laboratory; plus research and development agencies in West Germany and Sweden.

TAE is backed by a Goddard-based TAE Support Office, which assists users who have questions or specific problems. There is also an active user group that participates in defining new TAE development areas and organized users' conferences to exchange ideas and experiences. TAE is distributed through NASA's Computer Software Management and Information Center (COSMIC) at the University of Georgia.

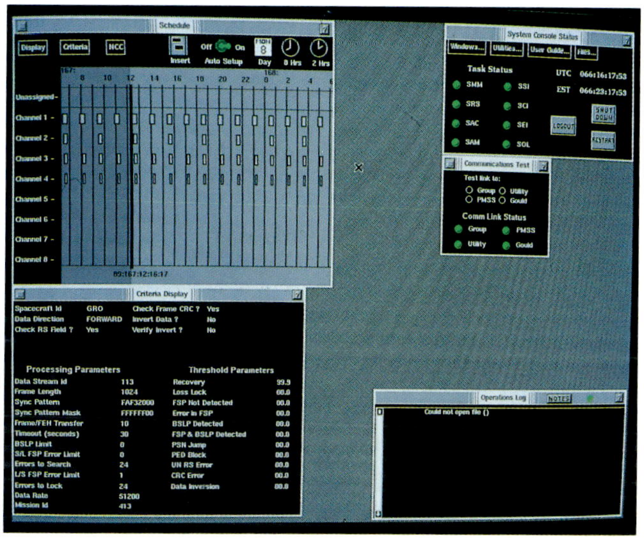

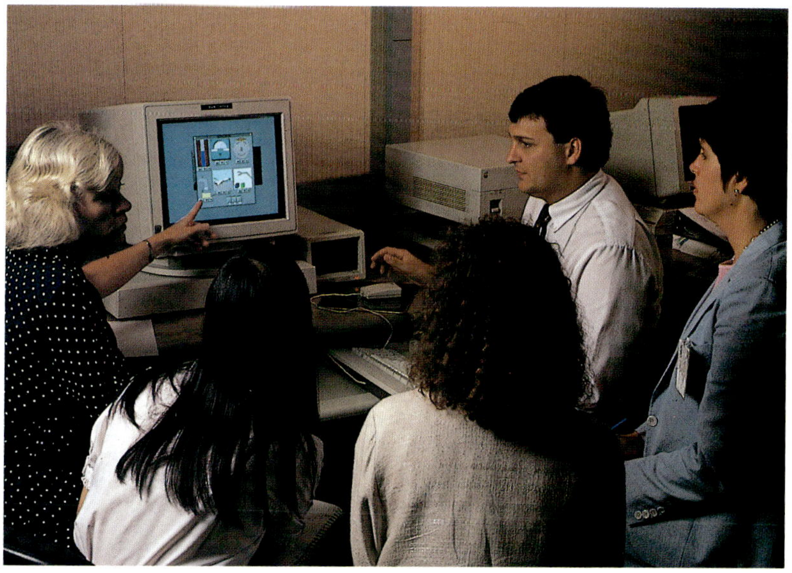

Error-free Software

Many computer engineers feel that error-free software, or even software close to error-free, is an impossible goal. Not Margaret Hamilton (right), president of Hamilton Technologies, Inc. (HTI), Cambridge, Massachusetts. A pioneer of ultrareliable software for more than 20 years, she believes it is possible to build ultrareliable software and her company is already offering a product that is capable of eliminating at least 75 percent of all software-related errors.

The product is known as 001™, described by HTI as "an integrated tool suite for automatically developing ultrareliable models, simulations and software systems." It has been applied in aerospace engineering, manufacturing, banking and software tools development.

Margaret Hamilton explains the 001 tool suite: "The essence of the 001 is that it provides the ability to simplify the complex. Its focus is to support the thinking process to be more reliable; it helps someone define his thoughts as simply as possible, but not simpler. The adherence to a philosophy of defining systems derived ultimately from a small set of reliable constructs is the reason why this approach is so powerful. 001 assumes that every system is a candidate for reuse. To build a reliable system, only reliable systems are used as building blocks and only reliable systems are used as mechanisms to integrate these building blocks. The new system becomes a reliable system for building larger systems."

A system developed with 001 can be prototypes or fully developed with production quality code. It is free of interface errors to a fine degree; it is consistent and logically complete and it has no side effects; errors of data and control flow, including those having to do with timing and priority, are eliminated. Says HTI: "Because of its features of reliability, automation, abstraction, flexibility and reusability, systems can be designed, developed and maintained with maximum productivity."

The 001 tool suite resides in the VAX/VMS environment but it can be ported to other machines. At left below, HTI software engineer Carl Kraenzel is working at a Sun System using the 001 tool for modeling; at right, HTI's Robert Poirier is creating a system on the Apple Mac 2.

The 001 tool suite represents a second development in Hamilton's quest for error-free software. She directed the research and development of an earlier product called USE.IT, the first computer-aided software engineering product in the industry, to concentrate on automatically supporting the development of an ultrareliable system throughout its life cycle. Both products, and the companies producing them, had their origins in NASA technology developed during the Apollo Lunar Landing Program.

For development of the onboard guidance, navigation and control system for the Apollo spacecraft, Johnson Space Center awarded a contract to the Charles Stark Draper Laboratory, Inc., Cambridge, Massachusetts. Hamilton directed development of Apollo onboard software; she led her staff in performing an extensive analysis of software errors and how they happen. This led to the formation of a theory—called Higher Order Software—that embodies control axioms for eliminating errors early. That work provided the beginning of 001.

In 1976, Hamilton left Draper to form Higher Order Software, Inc., also in Cambridge, and she served as the company's president and chief executive officer from 1976 to 1984. Hamilton founded HTI in 1986 to develop 001.

™ 001 is a trademark of Hamilton Technologies, Inc.

Data Acquisition Systems

KineticSystems Corporation, Lockport, Illinois develops and produces high speed CAMAC (Computer Automated Measurement and Control) data acquisition systems for science and industry. Some of the company's most advanced products resulted from a joint study and development program with Langley Research Center in the mid-1980s. The technology developed in that effort spurred widespread spinoff applications and made KineticSystems/ Langley joint recipients of a Research and Development Magazine IR-100 Award as one of the most significant advances of 1986.

The study involved feasibility determination of using CAMAC equipment to provide a distributed input/output system for Langley's Advanced Real Time Simulation (ARTS) system, which supports flight simulation R&D in such areas as automated control, navigation and guidance, air combat, and workload analysis for pilots and astronauts.

The study found CAMAC an ideal approach that would allow up to 32 high performance simulators located throughout the Langley complex to be controlled by centrally-located host computers. With Langley input, KineticSystems proceeded to develop the hardware for ARTS. Much of the CAMAC equipment was off-the-shelf, but the project demanded development of a new enhanced performance data highway and modules with higher resolution converters. KineticSystems developed a fiber optic highway that provided transmission of data between the simulators and the host computers at a rate of 24 million bits per second, allowing simulators in several locations to interact in real time. The company also developed a series of 16-bit analog to digital, digital to analog, and digital to synchro converter modules.

The technology developed in the ARTS project significantly boosted KineticSystems'

technical capability and brought a great variety of new applications in both the public and the private sectors. As cooperative marketing partners, KineticSystems and Digital Equipment Corporation, Marlborough, Massachusetts are delivering equipment derived from the ARTS work to almost 80 users in the U.S. and abroad. A typical CAMAC data acquisition system with operator interface is shown at far left; in the lower left photo is a CAMAC chassis and a sampling of the data acquisition and control modules available. The near left photo illustrates one of many applications, monitoring steelmaking operations.

The diversified range of applications includes such uses as fusion research, power grid analysis, process automation, turbine testing, petroleum distribution, chemical processing and steelmaking.

The equipment is being used in flight simulation research by Ames Research Center, Wright-Patterson Air Force Base and The Boeing Company. Major U.S. research laboratories, such as Argonne, Brookhaven, CalTech and Stanford, are using the CAMAC equipment for particle accelerator research. Princeton Plasma Physics Laboratory, Lawrence Livermore Laboratory and General Atomic are using the equipment in fusion research, as are foreign researchers in Japan and West Germany. Additionally, it is employed in high energy physics studies at Jawaharlal Nehru University in India, the CERN Research Laboratory in Switzerland, KEK National Laboratory in Japan, the University of Melbourne in Australia and KFA Laboratory in West Germany.

Among corporate users are AT&T, IBM, Standard Oil, TRW Inc., B.F. Goodrich, Hercules, Inc., Bonneville Power Authority, Corning Glass, General Electric, Koch Refining Company, Westinghouse, Whirlpool and LTV Steel.

A Spinoff From Mariner

Today only dimly remembered by space enthusiasts and not at all by the general public, the Mariner missions of the 1960s and 1970s constituted one of the most successful of all NASA programs in the early days of U.S. space research.

The Mariners were a family of planetary spacecraft for studies of Venus and Mars. There were nine of them; three failed during or shortly after launch, but the others scored spectacular successes for their day.

A space-derived dry lubricant leads a sampling of technology transfers in industrial productivity and manufacturing technology

Launched in 1962, Mariner 2 passed within 10,000 miles of Venus and became the first successful planetary probe. Mariner 4 (1964), a Mars explorer, was the first spacecraft to return pictures of another planet. Mariner 5 (1967) made the second U.S. flyby of Venus, closer this time at 2,500 miles from the planet's surface. Mariners 6 and 7 (1967-69) flew within 2,000 miles of Mars and provided some 200 pictures. Mariner 9, launched in 1971, went into orbit around Mars, sent thousands of photos of the Red Planet and its moons, and mapped one-third of the Martian surface.

The Mariners were developed in a period of rapidly advancing technology, so each model contained some improvements over its predecessor. The last four, Mariners 6 through 9, represented a second generation of the family, larger, heavier and considerably more sophisticated than the earlier spacecraft.

These latter Mariners incorporated a great deal of what was then considered leading edge technology—a variety of advances in onboard power, scientific instrumentation, communications and imaging/data transmission systems. Among all these improvements was an unsung technology: a dry film lubricant designed to meet the special needs of Mariner missions. Developed for NASA by Dr. Robert D. Nelson of Stanford University, it offered exceptional lubrication quality for reduced friction and extended wearlife of mating parts operating in harsh interplanetary environments where temperatures ranged from well below zero to 500 degrees Fahrenheit.

The technology was subsequently acquired and refined by Micro Surface Corporation, Morris, Illinois, which markets the lubricant as the WS2 modified tungsten disulfide coating. A pressurized refrigerated air application process impinges a dry metallic WS2 coating without heat, curing, binders or adhesives. The lubricant coating binds instantly to any metal or resin substrate with a thickness of 20 millionths of an inch.

In the aftermath of the Mariner missions, the dry film lubricant found its way into industry use, but only by aerospace and defense contractors. In 1984, Micro Surface introduced WS2 to general industry and it has since compiled an excellent performance record in an ever-widening range of applications among the automotive, medical equipment, plastics, tool and die, and robotics industries. It has been used, for example, to coat machine tools, industrial gears and bearings, electric motors, compressors, cryogenic pumps and small firearms.

In the plastics industry, WS2 users have found that in some operations, such as blow molding, injection molding and extrusions, the coating increases production by reducing the drag between tool steel and resin. In automotive applications, it is used to reduce friction and wear by Ford Motor Company, General Motors and Chrysler Corporation in such components as auto bearings, transmissions and engine internal parts. In special applications, it is used by racing hydroplanes cars of the "Indy," NASCAR and Winston Cup types.

In addition to reducing friction and wear, WS2 offers a number of other advantages, depending on the application; generally, it helps improve product quality, extends equipment service life and eliminates or reduces costly maintenance problems. It is finding growing acceptance and Micro Surface's list of WS2 customers reads like a Who's Who of American Industry. In addition to the U.S. automotive Big Three, a random selection includes American Can Corporation, Continental Can Corporation, Kimberly Clark Company, Dow Corning Corporation, Ethyl Corporation, General Electric Company, Phillips Petroleum, Whirlpool Corporation—and, of course, NASA.

In the top photo general manager Ed Fabiszak of Micro Surface Corporation displays a tool for making plastic parts that has been coated with WS2, a dry lubricant originally developed for space use which has found a wide range of practical Earth uses. A few of the many applications: in manufacture of plastic parts such as those shown above, companies coat injection molds to reduce sticking and increase production; automative companies use WS2 to lubricate a variety of parts, such as the pistons shown at lower left, to reduce friction and wear; in the robotics industry, bearings and sleeves (upper left) designed for repetitive movements are coated to extend their useful lives, as are the drill bits and milling tools (near left) used in heavy industry and machining.

Laser Technology

At right is a new green microlaser introduced last year by Amoco Laser Company, Naperville, Illinois. At bottom right is the same company's breakthrough infrared diode-pumped microlaser. Both are spinoff products based on a laser concept originally developed at NASA's Jet Propulsion Laboratory for optical communications over interplanetary distances.

A laser is a device that emits a narrow and very intense beam of light or other radiation; the beams may be employed, for example, to transmit communications signals, to drill, cut or melt hard materials, or, in medical applications, to remove diseased body tissue.

Microlasers are revolutionary new miniaturized all-solid-state lasers that cover a broad portion of the wavelength spectrum and offer dramatically improved performance over traditional lasers. Amoco Laser Company, a leader in microlaser technology, offers 20 different microlaser products for an expanding range of applications that includes medical instrumentation and therapy, color separation equipment for graphics and printing, film reading and writing, advanced projection TV, telecommunications and optical memory storage, plus a variety of industrial R&D/production applications such as micromaterials processing, spectroscopic and analytical measurement, semiconductor processing and characterizing optical and electronic materials to determine quality.

Amoco Laser, a wholly-owned subsidiary of Amoco Corporation, was formed to commercialize a variety of laser-based technologies developed by Amoco Technology Company's Research Department. In addition to Amoco-developed technology, the company has also acquired other patents relating to solid-state lasers pumped by tiny diodes, including the NASA technology in the infrared and green microlasers.

That technology was developed at Jet Propulsion Laboratory by Donald L. Sipes, Jr. of California Institute of Technology (CalTech). Subsequently, NASA waived the patent rights to CalTech and CalTech licensed the technology to Amoco Laser. According to Sipes, the patent centers on the discovery that a diverging, elliptically-shaped laser beam, such as is emitted by a laser diode, can be used to pump a solid-state laser very efficiently and also produce a very narrow, ideal laser beam. Inventor Sipes has since joined Amoco Laser as manager of product development.

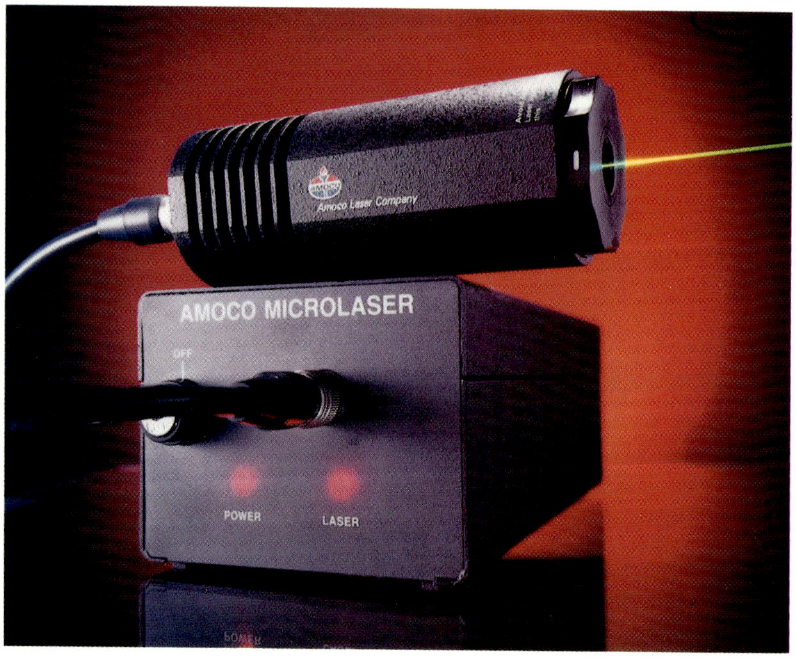

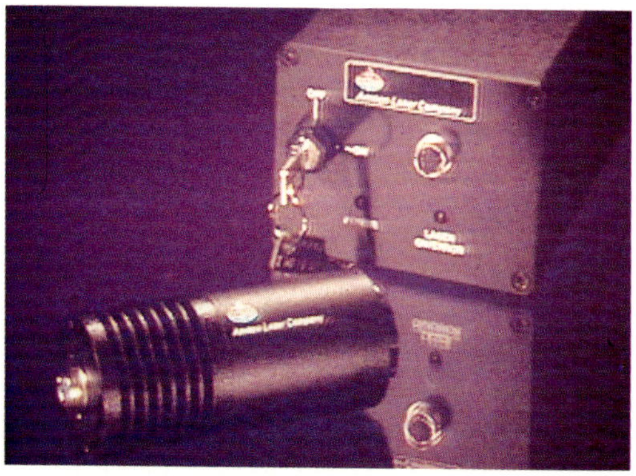

Inspection Tools

At right, a segment of a Space Shuttle Solid Rocket Motor is being scanned by a motorized contamination sensing device to assure surface cleanliness prior to bonding of a rubber liner.

The sensor and scanning system are part of a family of OP1000 Surface Quality Monitors developed by Photo Acoustic Technology Inc. (PAT), Westlake Village, California. The monitors are based on an inspection tool and technique known as Optically Stimulated Electron Emission (OSEE), invented in the early 1980s by Dr. Tennyson Smith of Rockwell International Corporation under contract to Marshall Space Flight Center.

PAT, which produces the sensors and scanning systems but not the associated robotics, was founded on this technology, which is described as a significant advance in measuring thin layer contamination on the surface of a material. PAT founders, including Mantosh Chawla, president and chief executive, learned of the technology through an article in Tech Briefs, a NASA publication that reports on new inventions or innovations developed in the course of NASA programs to let potential users know of technology available for transfer.

That beginning led to the development by PAT of the OP1000 line, a series of non-destructive, non-contact surface contamination detection systems with wide industrial applicability. Some examples: inspection of a variety of surfaces prior to bonding, coating, painting, plating, etching, soldering, brazing or welding; measurement of lubricant thickness in computer hard disks; examination of the surface chemistry of printed circuit boards; inspection of blank semiconductor wafers prior to processing; and on-line inspection of metal sheeting. The OP1000 series' realtime pre-processing detection capability assures 100 percent surface quality testing and reduces the need for destructive sample testing after the product process is completed.

The OSEE technique involves brief exposure of the surface to be inspected to high energy, low intensity ultraviolet radiation. The ultraviolet energy interacts with the surface layer, causing free electrons to be emitted from the surface; these emissions are picked up by the OP1000 system's detector.

Where contamination exists, in the form of very thin layer substances, it interferes with the electron flow and the degree of interference is proportionate to the thickness of the contaminant layer; this enables measurement of the thickness by means of system signal output that is proportional to the electron flow. A high output reading in the system's display window indicates a clean or acceptable surface; a low reading shows the presence of foreign matter and the level of contamination. OP1000 systems operate in a conventional atmosphere and do not require vacuum chambers or strict temperature control; they work on virtually any type of material and can detect either organic or inorganic contamination.

Manufacturing Aids

In the early 1980s, the Materials and Manufacturing Center (MMTC) of TRW Inc., which has since become MMTC/Textron, Cleveland, Ohio, conducted a research program for Lewis Research Center involving use of "thermal barrier" coatings to improve aircraft engine efficiency and reduce fuel consumption. The coatings, applied to turbine blades, combustors and other engine parts, allowed increasing the operating temperature of an engine by several hundred degrees, thus increasing overall efficiency.

For applying the coatings, MMTC invented a computer-aided, fully automatic robotic system for spraying a very hot plasma onto a turbine blade (right). MMTC also found it necessary to develop a means of controlling the thickness of the plasma deposit, which is measured in thousandths of an inch. This led to development of advanced optical gaging techniques to monitor and control point-to-point plasma spray deposition buildup on the surfaces of turbine blades.

These optical gaging techniques became the basis for a family of computerized optical gages built by MMTC for inspecting aircraft or industrial turbine and compressor blades, vanes and other parts of complex shape. MMTC/Textron offers 10 standard commercial robotic gages, including the Textron Model 501B shown at far right; it is in production use for in-process and final inspection of compressor blades and vanes. In addition to measuring and analyzing the airfoils in three dimensions, the system generates two dimensional profiles for assessing the status of—and specifying repairs to—the electrochemical machining cathodes used to make the parts. The Model 501B, and other similar systems, are capable of production floor accuracies to a ten-thousandth of an inch at speeds much faster than coordinate measuring machines.

MMTC/Textron is now marketing an expanded line of advanced optical gages for high precision, complex geometry applications on blades, vanes, integrally bladed rotors, impellers, integrally cast stators and other turbine parts. The gages employ laser, fiber optic or structured light video camera sensors as the application dictates. At lower right is one member of the new family, the Textron Model 6000.

These spinoff gages, says MMTC/Textron, not only offer multiple improvements in controlling the quality of manufactured parts, they also provide significant savings in the overall cost of the manufacturing process.

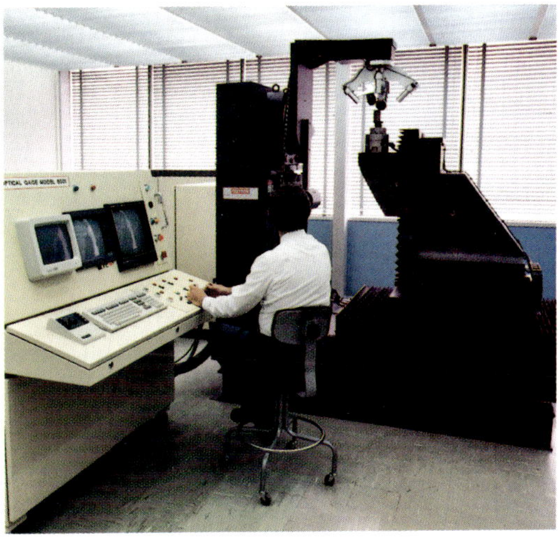

Wood Bond Testing

At right is a Bondcheck™ inspection system, a significant advance in production quality control that offers a nondestructive method of measuring the internal bond strength of wood particle board.

The internal bond analyzer (IBA) performs an acoustic-ultrasonic bonding efficiency measurement, combining ultrasonics with the acoustic emission testing technique to overcome shortcomings of either method and produce a highly efficient hybrid system. The IBA resulted from a multiyear joint development program between Hartford Steam Boiler Inspection Technologies, Sacramento, California, and The Weyerhaeuser Company, Tacoma, Washington, a major forest products firm. The system is thus a spinoff from an aerospace spinoff, stemming from a NASA-invented acousto-ultrasonic technique and test instrument that became a commercial system for testing the bond strength of composite materials.

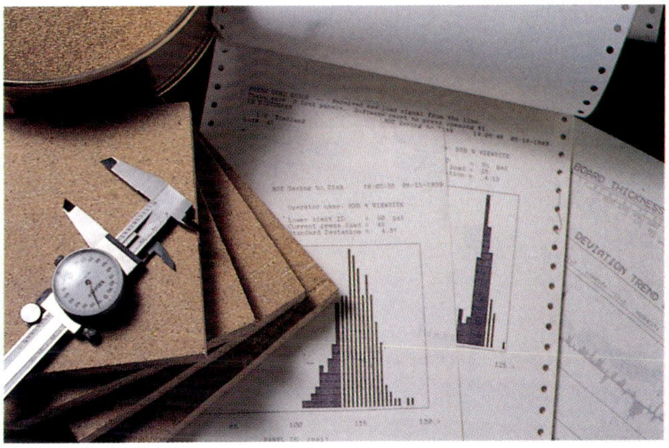

In the latter 1970s, Lewis Research Center was looking for better ways to test composites, which at that time were finding rapidly expanding acceptance in a variety of aerospace applications. Lewis wanted a nondestructive method that would not only detect flaws but also evaluate a composite material's strength and endurance. The center came up with the hybrid system that combined a proven metal-testing technique—ultrasonics—with the then relatively new technique known as acoustic emission testing.

In the original system, ultrasonic "stress waves" were injected into a composite structure. As the stress waves propagated through the material, the waveform was changed if flaws or irregularities existed. The changed signal from the ultrasonic pulser was detected by an acoustic emission sensor and analyzed. In a display section of the instrument, flaw and strength assessment information was presented in both graphic and digital form.

The Lewis technology was acquired by Acoustic Emission Technology Corporation (AET), a predecessor company of Hartford Steam Boiler Inspection. AET refined and commercialized the technology and produced, in the early 1980s, two types of acousto-ultrasonic test systems known as Model 206 and Model 301. The IBA builds on the AET technology and moves into a new area of application: on-line process control systems.

Previous methods of determining the internal bond strength of wood particle board involved destructive tests on parts cut from boards; the far left photo exemplifies the "old way" in which adhesives were applied to board samples and then the samples were pulled by machine until they came apart. There was a substantial time lapse before test results were available to the particle board mill operator.

The IBA system determines bond strength by measuring the changes in pulsed ultrasonic waves injected into a board by a specially designed rolling contact transducer (top). Analysis of the waveforms by a proprietary method of calculation determines the average internal bond strength for the panel. Results are displayed immediately, while the panel is still on the conveyor production line. Customers can get a computer printout (above) showing the quality consistency of their shipments.

The system offers additional advantages: the mill operator can adjust the proportion of resin to wood particles and reduce setup time required when changing to a different product. The IBA also offers startup waste reduction, consistent quality internal bonds and automatic marking of deficient products. These advantages add up to substantial cost savings.

™ Bondcheck is a trademark of Hartford Steam Boiler Inspection Technologies.

Cable Tester

Sonics Associates Inc., Birmingham, Alabama is a multifaceted corporation involved in design, manufacture and installation of professional audio-visual-video systems. Among other activities, Sonics supplies the audio—including several company-developed specialized components—for the giant Imax® and OMNIMAX® motion picture theaters on five continents (Sonics and Imax Systems, Toronto, Canada work closely together and Imax Systems recently acquired half ownership of Sonics). At right is the Imax theater at the Maryland Science Museum in Baltimore; a Sonics Associates engineer is checking a segment of the miles of ribbon cables running throughout the theater.

Sonics Associates' Imax and OMNIMAX Theater Sound Systems are required to deliver sound to complement motion picture images that may measure more than 70 feet high and 100 feet wide. Imax/OMNIMAX sound tracks are very dynamic and often simulate such dramatic effects as cannon fire, avalanches and rocket launches at sound levels approaching those of the actual event.

The production and installation of such sound systems demands a great many cables and connectors, and there are hundreds of separate conductors to be assembled and verified before connection to the equipment. This poses a problem: often the wiring must be done by local labor not as familiar with proper procedures as might be desired. Unless each cable connection can be verified, there is high risk of improper operation and damage to costly equipment.

Over the years, Sonics has sought to minimize the problem by employing a great variety of techniques and equipment for field testing interconnection cables; they range from ohmeters, lights, buzzers and telephones to sophisticated microprocessor-based testers. Each is generally effective but each has drawbacks. The simplest systems demand at least two people and considerable time, the more complex systems require fewer people but a disproportionate amount of expensive equipment.

NASA technology provided Sonics a means of saving countless hours of testing time. Ames Research Center had a similar problem: when computer communications cables are installed to connect remote terminals with a central

computer, errors frequently occur in installing multipin connectors and often the required functional signals are hooked up to wrong pin locations in the connectors. Such misconnections cause malfunction of the computer terminal, sometimes damage to the terminal or computer. To combat this problem, Ames developed a simple, two-part cable testing device that included an active part plugged into one end of the cable and a passive part at the other end.

Sonics vice president James B. Cawthon read about the Ames device in NASA Tech Briefs, a publication that describes technology available for transfer. He requested and received from NASA a Technical Support Package that provided extensive detail of the technology. Sonics made several changes to the design to adapt it to its own needs and came up with an inexpensive but highly effective tester that uses a clocked shift register to apply a voltage to a cable under test; this is the active part of the Ames development. The passive companion to the test generator is a small box containing light emitting diodes (LEDs). When connected to the other end of the cable being tested, the LEDs light in the same sequence as the generator. This simple procedure allows a technician to observe the sequence of the applied test voltage and note any discrepancy that would indicate a miswired cable.

The two units of the Sonics testing device are shown at top. At far right, Sonics' James Cawthon is using the device to check one of the control panel circuit boards at the Maryland Science Museum Imax theater to make sure nothing is cross-wired. At right center and in closeup at lower right, Cawthon is checking the synchronization of the film projector and the audio tape.

The NASA technology has saved Sonics a great deal of time and money, according to Cawthon, who adds that the tester can be built for about $25 and it takes minimal space in a tool box. "We haven't documented total time saved," he says, "but suffice to say that the device reduces the time to test a single cable from 12 minutes for two technicians to less than one minute for one technician."

® Imax and OMNIMAX are registered trademarks of Imax Systems Corporation.

Lubricant Coating Process

NASA and its contractors employ a wide variety of materials to reduce wear and sustain lubrication for aerospace vehicles operating in harsh environments. One such is molybdenum disulfide, a slippery gray-black powder used as a lubricant and a lubricant additive.

This material's unusual characteristics allow it to maintain lubricating ability at temperatures ranging from far below zero to more than 750 degrees Fahrenheit. Additionally, "moly" performs well under vacuum conditions and is only minimally affected by high radiation levels, hence has found extensive application in space systems. Moly is also in widespread use in non-aerospace industries, particularly in the automotive, railroad, airline, metal fabricating and petroleum products industries.

Like most solid lubricants, moly has a drawback: it must be perfectly bonded to a surface so that it does not "migrate" away from highly loaded contact areas or lose strength at elevated temperatures. To overcome that drawback, NASA developed a "peen plating" process for applying moly. Peening involves bombarding a surface with a high velocity stream of small shot, which act like thousands of tiny ball peen hammers pounding the surface. The pounding creates many microscopic surface indentations, thus forming tiny pockets of lubrication that prevent lubricant migration and promote high energy bonding; it also increases fatigue strength of the peened part and reduces stress corrosion.

This technology is now being employed under NASA license by Techniblast, Seminole, Oklahoma as the key element of the company's SURFGUARD process for applying high strength solid lubricants. Techniblast was assisted during the licensing process by Rural Enterprises, Inc., Durant, Oklahoma, a NASA application team.

Solid lubrication coatings are usually applied only in service centers due to process restraints. But Techniblast believes that most customers would prefer to have their own in-house coating capabilities, so Techniblast is extending the technology to include new equipment that can readily be used by industrial firms in their own facilities.

The SURFGUARD process requires two machines—one for cleaning and one for coating. The cleaning step is necessary so that the coating is bonded directly to the substrate to provide a better "anchor" for the coating. The coating machine, designed to handle moly and other powdered lubricants, applies a coating half a micron thick. A blast gun, which can use various pressures to vary peening intensities for different applications, fires high velocity "media"—the peening hammers—ranging from plastic pellets to steel shot.

At left is a Techniblast-developed prototype coating machine. At top is an interior view of the machine showing a filtering unit that recaptures the airborne powder for reuse. At right is the coating nozzle. The lower photos are views of an aluminum die cast mold before (right) and after application of SURFGUARD. While developing SURFGUARD equipment for in-house use by customers. Techniblast is also offering coating service performed at its own facility.

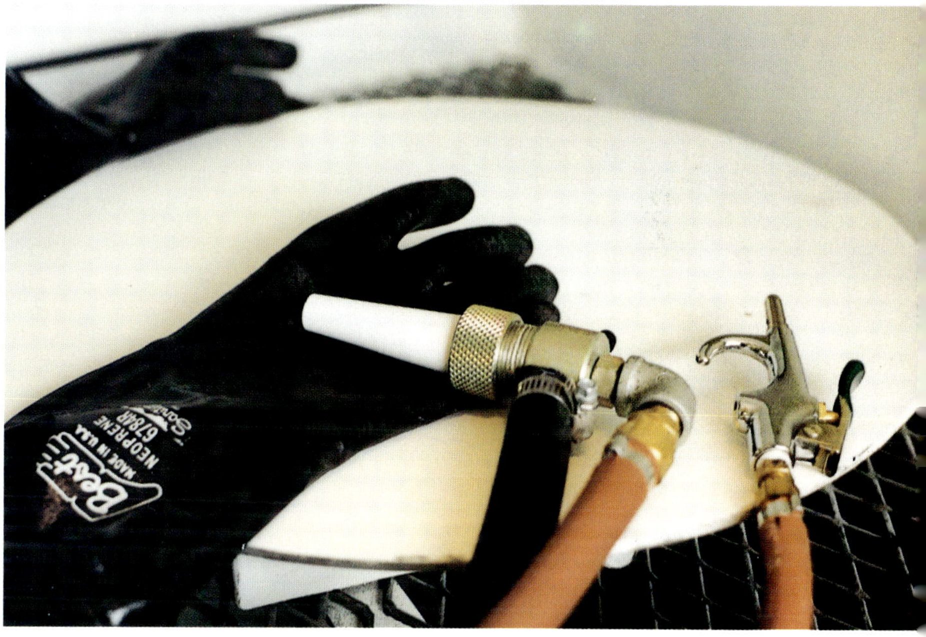

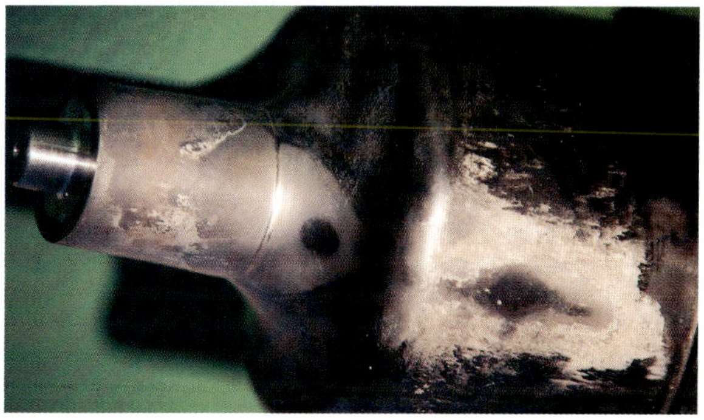

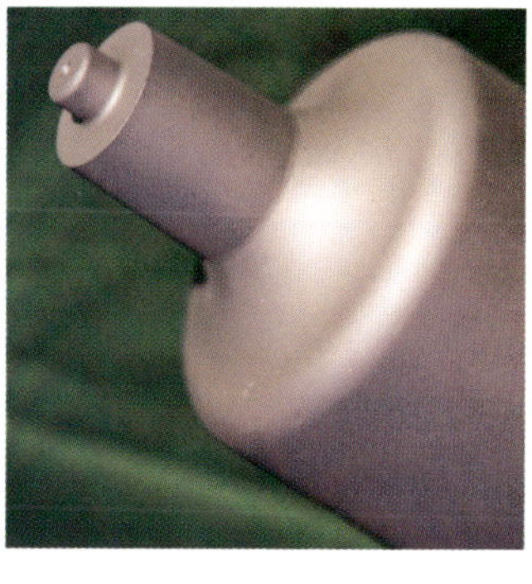

Explosive Joining

To permit metal joining operations under hazardous or inaccessible conditions—for example, joining structures in space or repairing nuclear reactors—Langley Research Center developed a small scale explosive seam welding process that allows remote joining and provides a joint with double the strength of the parent metal.

The basic technique, invented by Langley's Laurence J. Bement, involves use of very small quantities of ribbon explosive to create hermetically sealed joints. Joining is accomplished by the collision of the metal plates to be joined as they are slammed together by the force of the explosion. This collision causes a skin deep melt and ejection of oxide films on the surfaces being joined, allowing a linkup of electrons that produces superstrong, uniform joints.

The explosive technique can be used to join materials that otherwise would not join, including a wide variety of metals, alloys and combinations. It offers a number of advantages over mechanical fasteners, adhesives and other types of joining, and it has a wide range of potential applications.

Langley Research Center and inventor Bement provides technology transfer assistance to Demex International Ltd., Picayune, Mississippi, which has refined the NASA technology and adapted it to commercial applications, such as plugging leaking tubes in feedwater heaters (right and right below) used by utility companies. Demex produces the small plugs, their associated sleeves and detonators shown at top right. At center right is a closeup view of a plug being inserted in a feedwater heater tube sheet. At far right below is a microphoto of a plug/tube weld showing a "textbook" wave form that indicates a perfect weld.

Demex International teams with Southwestern Engineering Service Company, King of Prussia, Pennsylvania, which does the service work; together they have made some 35,000 plug installations without an operational failure. The explosive welding technology, according to Southwestern Engineering, allows faster plugging, hence reduced downtime, cuts plugging costs and increases reliability.

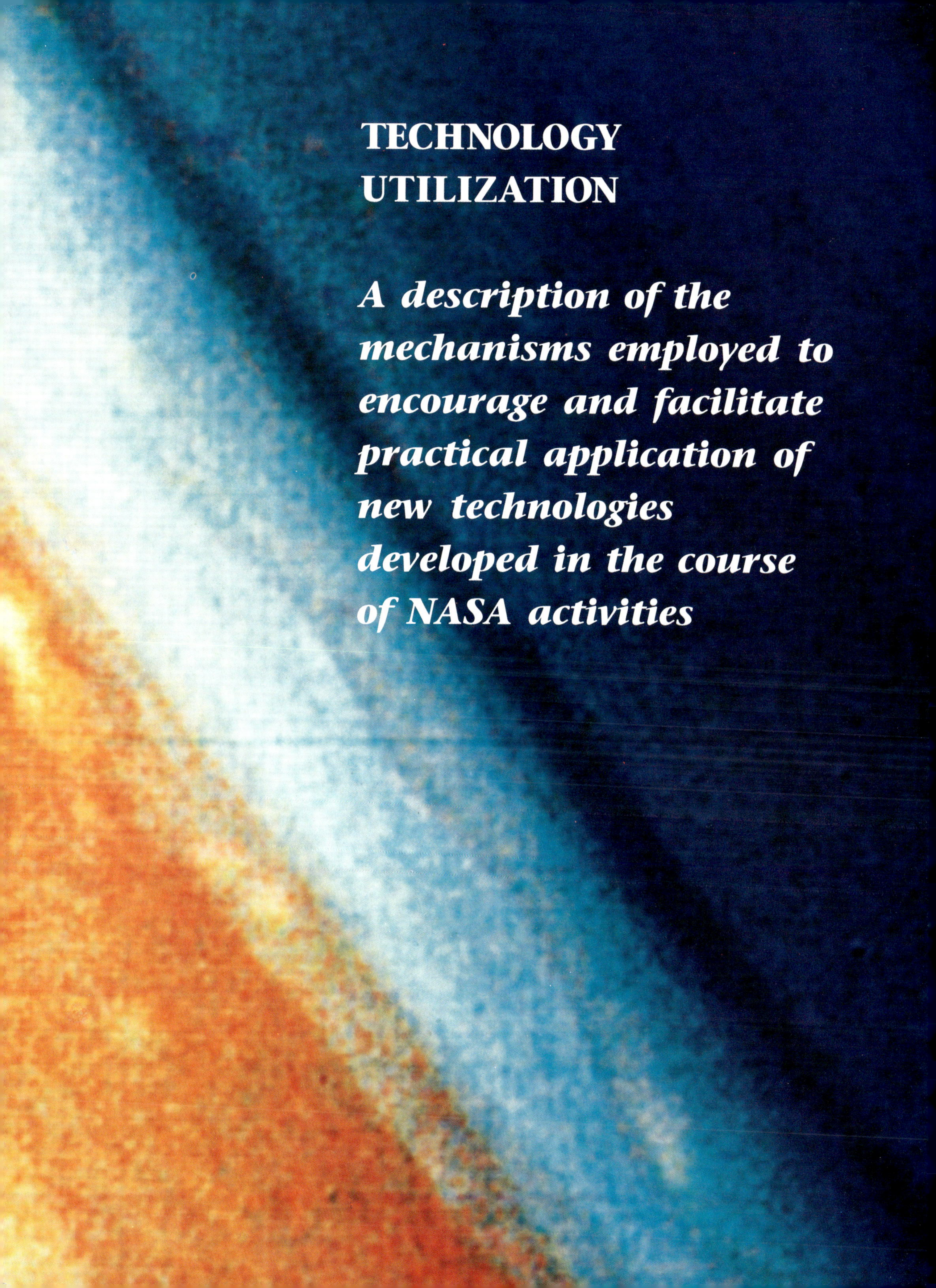

TECHNOLOGY UTILIZATION

A description of the mechanisms employed to encourage and facilitate practical application of new technologies developed in the course of NASA activities

Putting Technology To Work

By their challenging nature, NASA programs are especially productive of advanced technology. This wealth of technology is an important national asset in that it can be reused to develop new products and processes to the benefit of the U.S. economy in expanded productivity. But such secondary applications do not happen automatically; it takes an organized effort to put the technology to work in new ways and to reap thereby a dividend on the national investment in aerospace research.

A nationwide technology utilization network seeks to broaden and accelerate secondary application of NASA technology in the public interest

NASA's instrument for accomplishing that objective is the Technology Utilization Program, which employs several types of mechanisms to stimulate the transfer of aerospace technology to other sectors of the economy. The program is managed by the Technology Utilization Division, a component of NASA's Office of Commercial Programs. Headquartered in Washington, D.C., the division coordinates the activities of technology transfer specialists located throughout the United States.

A key element of the program is a network of 10 Industrial Applications Centers (IACs) affiliated with universities across the nation. The IACs offer clients access to a great national data bank that includes some 100 million documents of accumulated technical knowledge, along with their expertise in retrieving information and applying it to clients' needs. The IACs are backed by more than a score of Industrial Applications Center Affiliates, state-sponsored business or technical centers that provide access to the NASA technology transfer system.

A representative IAC is the NASA Industrial Applications Center (NIAC) at the University of Pittsburgh, Pennsylvania. NIAC's purpose is to provide U.S. industry and individual entrepreneurs access to existing and evolving technologies, with the aim of enhancing their innovation and productivity. The center has helped thousands of companies to find new applications; to stay abreast of scientific, technical and business developments; to gain competitive intelligence; to identify qualified technical experts; and to monitor patent activity.

Among NIAC's technology assistance programs are document

retrieval; a problem-solving service; and a current awareness/update service that alerts a participant, throughout the year, to new developments in particular subject or research areas.

An example of how NIAC helps industry is found in the experience of Joseph A. Resnick, president of Dynamed Audio Inc., Natrona Heights, Pennsylvania. Resnick invented and introduced to the commercial market such products as a device that provides visual cues to deaf and hearing impaired people as a means of helping them achieve better speech; a "tone emitter," a miniature electronic voice box that enables a person who has lost his natural larynx to shape words; and a water pollution detection and alert system that monitors purity levels in home water filters and sounds an alarm when foreign substances are detected. NIAC helped Resnick with computer search and retrieval services on each of these inventions and a number of others.

Resnick calls NIAC his "research partner." "Before I commit any funds to developing an idea," he says, "I commission a complete technical and field search through NIAC. On the basis of the information they extract, I then determine whether or not my idea is feasible, and if feasible I commit and put the rest of the gears into play. I have developed a number of technologies, but had it not been for the accuracy of the information systems of NASA, I doubt seriously whether the devices would have come to fruition."

In addition to the IACs, other mechanisms of the Technology Utilization Program include Technology Utilization Officers, located at NASA field centers, who serve as regional managers for the program; a software center that provides, to industry and government clients, computer programs applicable to secondary use; applications engineering projects, efforts to solve public sector problems through the application of pertinent aerospace technology; and publications that inform potential users of technologies available for transfer. These mechanisms are amplified in the following pages.

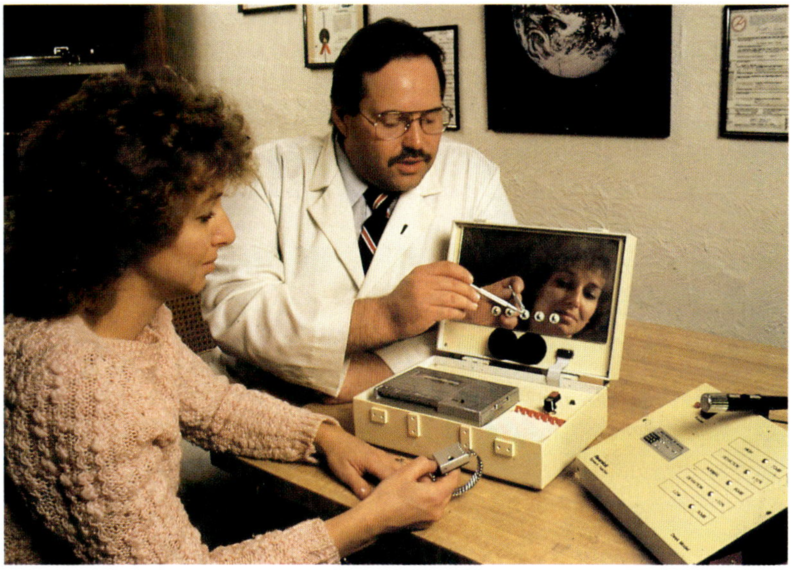

Inventor Joseph A. Resnick is describing the operation of his visual cueing speech improvement device for the deaf and hearing impaired.

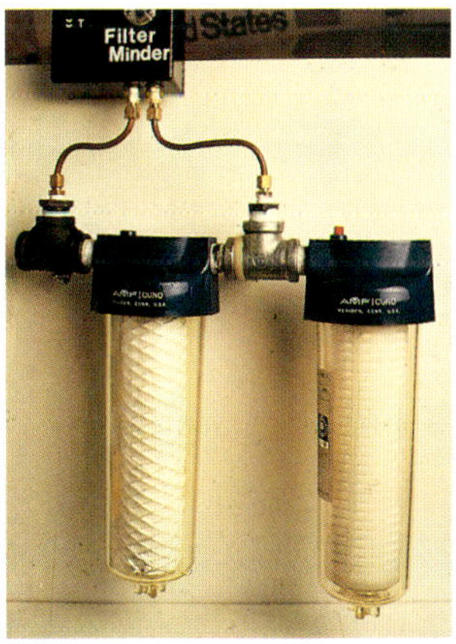

Another Resnick invention is a water pollution alert system that monitors purity levels in home water filters and sounds an alarm when foreign substances are detected.

Resnick is a regular client of the NASA Industrial Applications Center (NIAC) at the University of Pittsburgh and he credits NIAC's search and retrieval services with a big assist in bringing several of his inventions to the commercial marketplace. Resnick poses with his NIAC contact, Robert W. Baird.

Technology Utilization Officers

An important element among the NASA mechanisms for accelerating and broadening aerospace technology transfer is the Technology Utilization Officer or TUO. TUOs are technology transfer experts at each of NASA's nine field centers and one specialized facility who serve as regional managers for the Technology Utilization Program.

Representative of the group is Dean C. Glenn, TUO at Johnson Space Center (right). Glenn (dark jacket) is shown in the lower photo discussing an applications engineering project with his staff. Pictured with him are Dennis Morrison (standing, next to Glenn), Norwood Hunter (seated, wearing headset), Carol Nash (white blouse) and Donna Maloy Brown (seated).

The Johnson Space Center TUO is responsible for coordination and management of all applications engineering projects at the center. In 1989, Glenn's office was coordinating projects in engineering, life sciences, artificial intelligence, education and communications. The Johnson TUO is also responsible for maintaining awareness of NASA's technology transfer policies. In 1989, the TUO initiated outreach activities aimed at the adjacent Houston (Texas) community and focused on technology transfer and the Technology Utilization Program.

The TUO's basic responsibility is to stay abreast of research and development activities at his center that have significant potential for generating transferrable technology. He assures that the center's professional people identify, document and report new technology developed in the center's laboratories and, together with other center personnel, he monitors the center's R&D contracts to see that NASA contractors similarly document and report new technology, as is required by law. This technology, whether developed in house or by contractors, becomes part of the NASA bank of technical knowledge that is available for secondary application.

To advise potential users of the technology's availability, the TUO evaluates and processes selected new technology reports for announcement in NASA publications and other dissemination media. Prospective users are informed that more detailed information is available in the form of a Technical Support Package.

The TUO also serves as a point of liaison among industry representatives and personnel of his center, and between center personnel and others involved in applications engineering projects, which are efforts to solve public sector problems through the application of pertinent aerospace technology. On such projects, the TUO prepares and coordinates applications engineering proposals for joint funding and participation by federal agencies and industrial firms.

NASA conducts, independently or in cooperation with other organizations, a series of conferences, seminars and workshops designed to encourage broader private sector participation in the technology transfer process and to make private companies aware of the NASA technologies that hold promise for commercialization. The TUO plays a prominent part in this aspect of the program. He arranges and coordinates his center's activities relative to the meetings and when—as frequently happens—industry participants seek to follow up with visits to the center, he serves as the contact point.

Support for the TUOs—and for all the other elements of the NASA technology utilization network—is provided by the Technology Utilization Office at the NASA Scientific and Technical Information Facility (STIF). This office executes a wide variety of tasks, among them maintenance of the subscription list for *Tech Briefs* (below), NASA's principal tool for advising potential users of technologies available for transfer; maintenance and mailout of Technical Support Packages, which requires a reproduction effort of more than 1.5 million pages annually; and responding to requests for information, an activity that entails processing of some 120,000 letters and other inquiries and mailout of more than 200,000 documents yearly.

The TUO/STIF is also responsible for research, analysis and other work associated with this annual *Spinoff* volume; for distribution of technology utilization publications; for retrieval of technical information and referral of highly technical requests to appropriate offices; for developing reference and bibliographic data; and for public relations activities connected with media, industry and trade show interest in technology utilization matters.

Software Center

In the course of its varied activities, NASA makes extensive use of computer programs in such operations as launch control, analyzing data from spacecraft, conducting aeronautical design analyses, operating numerically controlled machinery and performing routine business or project management functions.

To meet such software requirements, NASA and other technology generating agencies of the government have of necessity developed many types of computer programs. They constitute a valuable resource available for reuse. Much of this software is directly applicable to secondary applications with little or no modification; most of it can be adapted to special purposes at far less than the cost of developing a new program.

Therefore, American businesses can save a lot of time and money by taking advantage of a special NASA technology utilization service that offers software capable of being adapted to new uses. NASA's mechanism for making such programs available to the private sector is the Computer Software Management and Information Center (COSMIC)®.

Located at the University of Georgia, COSMIC gets a continual flow of government developed software and identifies those programs that can be adapted to secondary usage. The center's library contains more than 1,400 programs for such tasks as structural analysis, design of fluid systems, electronic circuit design, chemical analyses, determination of building energy requirements, and a great variety of other functions. COSMIC customers can purchase a program for a fraction of its original cost and get a return many times the investment, even when the cost of adapting the program to a new use is included.

An example of how this service aids industry is the use of a NASA-developed program by Advanced Fuels Technology (AFT), Cleveland,

Ohio, a company engaged in developing environmentally acceptable forms of coal as an alternative to oil. AFT developed an economical process for cleaning coal that removes up to 85 percent of the sulphur and 75 percent of the ash. Coal thus cleaned provides more energy with less variability in burning characteristics and reduced emissions of air pollutants. At left is the control room of AFT's plant at Bridgeport, New Jersey where coal development work is conducted; at lower left is a ball mill that grinds the coal before cleaning.

In the course of their coal-cleaning technology development, AFT engineers tested unclean slurry against clean slurry to determine the amount of sulphur and ash that would be allowable. To evaluate the products left after combustion—solid, liquid and gaseous materials—AFT employed a computer program originally developed by NASA's Lewis Research Center to provide specific capabilities for determining products of combustion. AFT reported a saving of four man-months that would have been required to develop similar software had the COSMIC program not been available.

Another example involves a company that is a "double barreled spinoff" because it was founded to commercialize NASA data processing technology and, in addition, a major company development is an adaptation of a NASA-developed computer program. The company is Delta Data Systems, Inc. (DDS), Picayune, Mississippi, founded by a group of former NASA/industry engineers to commercialize remote sensing technology. At right, a DDS technician is computer enhancing an image developed from data supplied by the Landsat remote sensing satellite.

DDS used a COSMIC-supplied NASA program for processing remotely sensed data as a "shell" for developing the company's own ATLAS geographic information system, used to process satellite and aircraft data, to digitize soil and topographic maps and to generate land use maps. The NASA program is known as ELAS (Earth Resources Laboratory Applications Software). DDS officials estimate that use of ELAS as a basis for ATLAS development saved four man years that would have been required to develop the 100 applications modules in the ATLAS system.

To assist prospective customers in locating potentially useful software, COSMIC publishes an annual indexed catalog of all the programs in the center's inventory. Available on microfiche, computer magnetic print tape or in hard copy form, the catalog may be purchased directly from COSMIC. The center also helps customers define their needs and suggests programs that might be applicable. For further information about COSMIC's services, contact the director at the address in the directory that follows.

® COSMIC is a registered trademark of the National Aeronautics and Space Administration

Technology Applications

One facet of NASA's Technology Utilization Program is an applications engineering effort involving use of NASA expertise to redesign and reengineer existing aerospace technology for the solution of problems encountered by federal agencies, other public sector institutions or private organizations.

Applications engineering projects originate in various ways. Some stem from requests for assistance from other government agencies, others are generated by NASA technologists who perceive possible solutions to problems by adapting NASA technology to the need. NASA employs an applications team composed of several scientists and engineers representing different areas of expertise. The team members contact public sector agencies, medical institutions, industry representatives, trade and professional groups to uncover problems that might be susceptible to solution through application of aerospace technology.

An example of an ongoing applications engineering project is one that seeks to improve techniques for treatment of serious burns and reduce the death rate among burn victims.

About two million Americans suffer serious burns each year and some 200-300,000 of them require hospitalization. Among the latter, 70,000 receive intensive care and 10-12,000 die from their injuries.

The traditional treatment is to allow natural sloughing of burn-caused dead tissue, then to close the resulting open wounds with skin grafts. But natural sloughing may take weeks and often results in infection and sepsis; in fact, bacterial infection is the major cause of death among burn victims. Effective treatment, therefore, is dependent upon early recognition of the extent of dead tissue and its removal, by chemical or surgical means, to minimize risk of infection and hasten healing. The key is accurate information on the depth of burn—but current methods are largely subjective and thus prone to error.

To meet the need for precise determination of the degree of burn, NASA initiated an

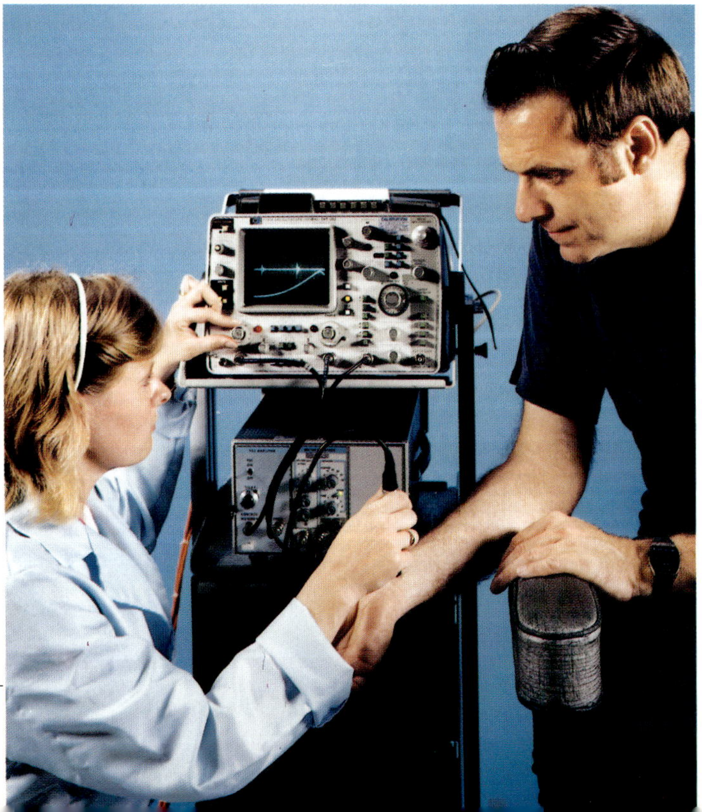

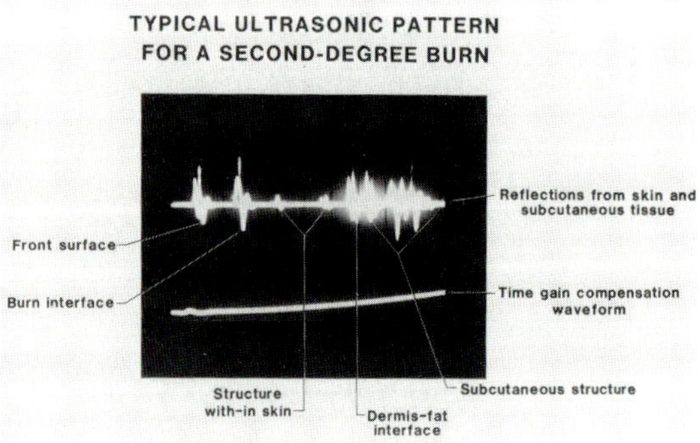

applications engineering project based on ultrasonic technology developed at Langley Research Center to detect flaws in metal components of aerospace systems. Langley developed an ultrasonic burn measurement instrument, using a combination of commercially available and specially designed components. Other organizations cooperating on the project include the Burn Unit of the Medical College of Virginia (MCV), Richmond, Virginia; the U.S. Army Institute of Surgical Research, Fort Sam Houston, Texas; and the NASA Technology Applications Team, Research Triangle Institute (RTI), North Carolina, which is coordinating the project and directing the effort to commercialize the technology.

The instrument directs ultrasound waves at the burn area; the difference in tissue density between damaged and healthy tissue causes the sound waves to reflect at the point of interface, and analysis of the reflections permits an accurate determination of burn depth. At far left, Langley's Dr. William T. Yost and Pamela Hanna-Hawver are operating a prototype model of the instrument; at left center is a typical ultrasonic pattern for a second degree burn as it appears on the instrument's screen.

The system was first tested with animals and test results showed that signals could be obtained that correlated well with microscopic analyses of tissue. Clinical tests by MCV on patients with different degrees and types of burns indicated that the technique can resolve second from third degree burns and that the type of burn—flame, scald, chemical—has little effect on the reflective characteristics of the burn. The clinical tests have been completed successfully and the RTI team is assisting in negotiating an agreement with Fulmer Dyson, Inc. for commercialization of the burn measurement instrument.

Another applications engineering project deals with development of a system to help memory-impaired "wanderers." Wandering behavior, frequently observed in older people suffering from Alzheimer's Disease and other mentally debilitating illnesses, can be a serious, sometimes life threatening problem for the disoriented person who wanders into a busy street or strays from home in severe weather. Wandering behavior is also extremely stressful and disruptive to members of the patient's family.

A corrective measure, a Locator System for Wandering Individuals development project, was initiated by Johnson Space Center (JSC) with support from the RTI Technology Applications Team; it is an interagency effort jointly funded by NASA, the Administration on Aging, the Veterans Administration, the National Institute on Aging, and the National Institute of Disability and Rehabilitation Research.

For the home setting, no device is available that will allow the wanderer freedom of movement yet warn the family caregiver of impending danger. Systems used in nursing homes have not proved totally effective and one study showed that only five percent of 170 nursing homes studied were using such systems. The JSC effort involves development of a system incorporating advanced tracking and communications technologies into a modular design that will monitor the patient's movement within a predetermined safe perimeter and send warning signals to both caregiver and wanderer when the patient begins to move beyond prescribed limits.

Cortex Electronics, San Bernardino, California, selected as the collaborating manufacturer, is working on prototypes of the wanderer notification device. Upon completion of the prototypes early in 1990, testing will get under way under the supervision of the cofunding agencies.

Publications

An essential measure in promoting greater use of NASA technology is letting potential users know what NASA-developed technology is available for transfer. This is accomplished primarily through the publication NASA *Tech Briefs*.

The National Aeronautics and Space Act requires that NASA contractors furnish written reports containing technical information about inventions, improvements or innovations developed in the course of work for NASA. Those reports provide the input for *Tech Briefs*. Issued monthly, the publication is a current awareness medium and problem solving tool for more than 150,000 government and industry readers. It is a joint publishing venture of NASA and Associated Business Publications of New York City.

Each issue contains information on newly developed products and processes, advances in basic and applied research, improvements in shop and laboratory techniques, new sources of technical data and computer programs, and other innovations originating at NASA field centers or at the facilities of NASA contractors. Firms interested in a particular innovation may get more detailed information by requesting a Technical Support Package; more than 120,000 such requests are generated annually.

Here are some examples of how *Tech Briefs* spreads the word and inspires secondary usage of NASA technology:

Experiments at Langley Research Center showed that small, barely-visible grooves on the surface of an airplane can reduce skin friction by as much as 10 percent compared with ungrooved surfaces. Called riblets, the grooves were machined into flat aluminum samples and tested in a wind tunnel. An article in *Tech Briefs* described the Langley work and riblets' potential for improving aircraft fuel efficiency by reducing drag.

Engineers at 3M Company, St. Paul, Minnesota contacted Langley with a suggestion: it would be simpler to mold grooves into a lightweight plastic film with adhesive backing and press the film into place on an airplane, allowing extension of the technology to existing airplanes as well as new ones. 3M and Langley cooperated on development and test of film riblets and found they were able to reduce drag as well or better than machined aluminum surfaces.

3M then reported the test results to The Boeing Company, the world's largest manufacturer of commercial transport aircraft, which

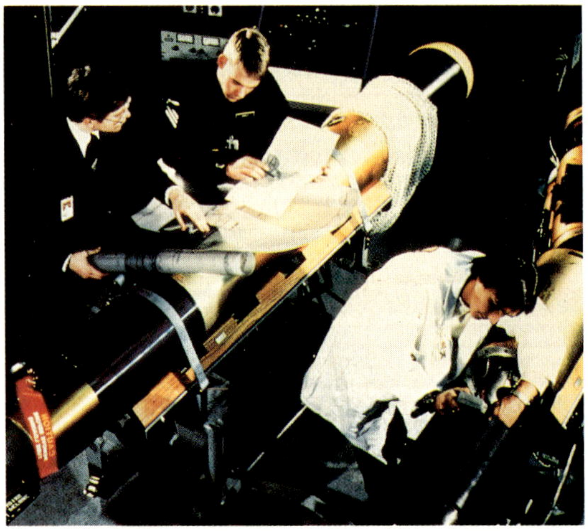

initiated its own research program investigating the efficacy of film riblets and their potential for airline fuel reduction. Doug McLean, a Boeing engineer, sparked another investigation, the applicability of riblets to water craft, specifically rowing shells (far left). So equipped, the U.S. men's four-oar-with-coxswain shell won an Olympic silver medal in an event wherein no U.S. team had won a medal for many years. Later, riblets helped the *Stars & Stripes* bring the America's Cup back to the United States.

Another example: engineers of Honeywell Inc. had a problem with the Mark 50 torpedos they were producing for the U.S. Navy (left center). Test runs disclosed that, as operating depth increased, ocean pressure tended to force sea water through the hull assembly joints, degrading torpedo reliability by possible short-circuiting of the electronic controls.

Honeywell sought a way of correcting the problem without expensive and time-consuming redesign of the whole system. A company engineer read in *Tech Briefs* of a superabsorbent fabric developed by Johnson Space Center (JSC) for capturing human wastes in manned spacecraft. Honeywell contacted JSC, procured the material and fabricated it into special containment devices on the Mark 50. The fabric can sequester up to 400 times its own weight in water, thus a relatively small amount of it can protect the Mark 50 from deepwater hull seepages. Instead of the great expense the company might have incurred in redesign, Honeywell's cost of correcting the problem was on the order of one man week.

Another example: *Tech Briefs* provided a productivity improvement benefit to Deposition Technology, Inc. (DTI), San Diego, California, information about a special type of valve—originally developed by Lewis Research Center—that relieves pressure beyond a specified limit by allowing gas to escape from a pressurized system. DTI installed the valves on two vacuum chambers that are part of the company's system for "sputtering," the process of applying filmlike metal coatings onto a surface by bombarding the coating material with electrocharged ions; this causes the material to distintegrate and relocate on the surface one layer at a time. Shown above left is a DTI sputtering chamber.

It is common practice to vent a vacuum chamber to a dry, inert gas following a chamber run. Because the gas is under pressure, it is necessary to protect delicate vacuum gauges and other components from overpressure. DTI's use of the NASA-developed valve (above right) not only protects the equipment but also frees the operator from monitoring the chamber venting. Fabrication and installation of the valves involved only nominal cost. DTI reported that the valves save an estimated 40 man-hours yearly in addition to avoidance of the substantial downtime and expense that might have resulted from overpressure of sensitive components.

Available to scientists, engineers, business executives and other qualified technology transfer agents in industry or government, *Tech Briefs* is the principal publication of the Technology Utilization Program. Among others are this annual *Spinoff* volume and the *NASA Patent Abstracts Bibliography*, a semiannually updated compendium of NASA patented inventions available for licensing, which now number almost 4,000 (the latter publication can be obtained from the National Technical Information Service, Springfield, Virginia 22101).

NASA's Technology Transfer System

The NASA system of technology transfer personnel and facilities extends from coast to coast and provides geographical coverage of the nation's primary industrial concentrations, together with regional coverage of state and local governments engaged in transfer activities. For specific information concerning the activities described below, contact the appropriate technology utilization personnel at the addresses listed.

For information of a general nature about the Technology Utilization program, address inquiries to the Manager, Technology Utilization Office NASA Scientific and Technical Information Facility, Post Office Box 8757, Baltimore, Maryland 21240.

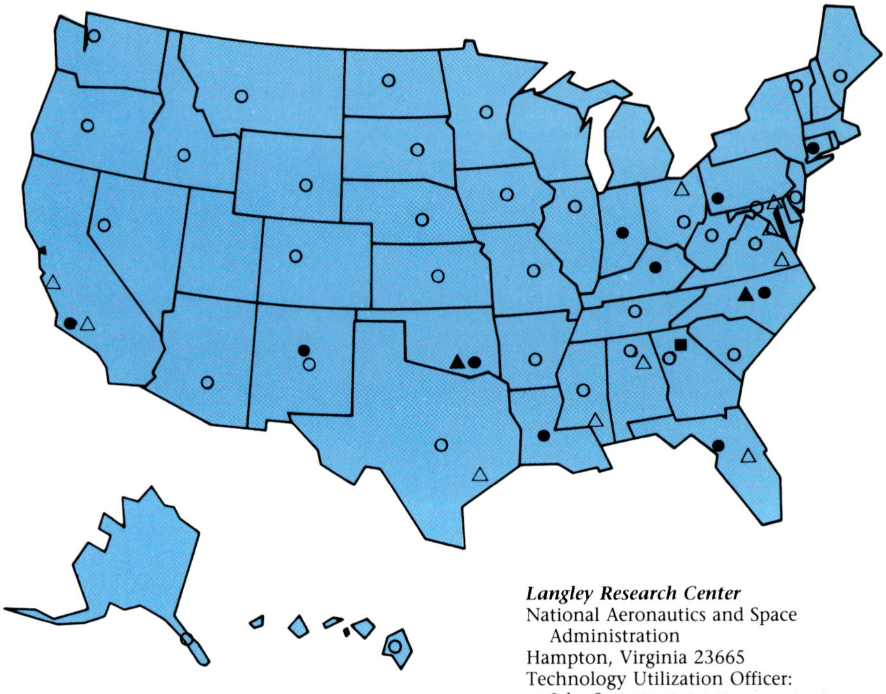

△ ***Field Center Technology Utilization Officers:*** manage center participation in regional technology utilization activities.

● ***Industrial Applications Centers:*** information retrieval services and assistance in applying technical information relevant to user needs.

○ ***Industrial Applications Center Affiliates:*** state-sponsored business or technical assistance centers that provide access to NASA's technology transfer network.

■ ***The Computer Software Management and Information Center (COSMIC):*** offers government-developed computer programs adaptable to secondary use.

▲ ***Application Teams:*** assist agencies and private institutions in applying aerospace technology to solution of public problems.

△ Field Centers

Ames Research Center
National Aeronautics and Space Administration
Moffet Field, California 94035
Technology Utilization Officer:
Laurence A. Milov
Phone: (415) 694-4044

Goddard Space Flight Center
National Aeronautics and Space Administration
Greenbelt, Maryland 20771
Technology Utilization Officer:
Donald S. Friedman
Phone: (301) 286-6242

Lyndon B. Johnson Space Center
National Aeronautics and Space Administration
Houston, Texas 77058
Technology Utilization Officer:
Dean C. Glenn
Phone: (713) 483-3809

John F. Kennedy Space Center
National Aeronautics and Space Administration
Kennedy Space Center, Florida 32899
Technology Utilization Officer:
Thomas M. Hammond
Phone: (407) 867-3017

Langley Research Center
National Aeronautics and Space Administration
Hampton, Virginia 23665
Technology Utilization Officer:
John Samos
Phone: (804) 864-2484

Lewis Research Center
National Aeronautics and Space Administration
21000 Brookpark Road
Cleveland, Ohio 44135
Technology Utilization Officer:
Harvey Schwartz (acting)
Phone:(216) 433-5567

George C. Marshall Space Flight Center
National Aeronautics and Space Administration
Marshall Space Flight Center, Alabama 35812
Director, Technology Utilization Office:
Ismail Akbay
Phone: (205) 544-2223

Jet Propulsion Laboratory
4800 Oak Grove Drive
Pasadena, California 91109
Technology Utilization Manager:
Norman L. Chalfin
Phone: (818) 354-2240

NASA Resident Office—JPL
4800 Oak Grove Drive
Pasadena, California 91109
Technology Utilization Officer:
Gordon S. Chapman
Phone: (818) 354-4849

John C. Stennis Space Center
Mississippi 39529
Technology Utilization Officer:
Robert M. Barlow
Phone: (601) 688-1929

● Industrial Application Centers

Aerospace Research Applications Center
611 N. Capitol Avenue
Indianapolis, Indiana 46204
F.T. Janis, Ph.D., director
Phone: (317) 262-5036

Central Industrial Applications Center
Rural Enterprises, Inc.
P.O. Box 1335
Durant, Oklahoma 74702
Dickie Deel, Ph.D., director
Phone: (405) 924-6822

NASA Industrial Applications Center
823 William Pitt Union
Pittsburgh, Pennsylvania 15260
Paul A. McWilliams, Ph.D., executive director
Phone: (412) 648-7000

NASA Industrial Applications Center
Research Annex, Room 200
University of Southern California
3716 South Hope Street
Los Angeles, California 90007
Radford G. King, director
Phone: (213) 743-8988
(800) 642-2872 (CA only)
(800) 872-7477 (toll free, US)

NERAC, Inc.
One Technology Drive
Tolland, Connecticut 06084
Daniel Wilde, Ph.D., president
Phone: (203) 872-7000

North Carolina Science and Technology Research Center
P.O. Box 12235
Research Triangle Park,
North Carolina 27709
H. Lynn Reese, director
Phone: (919) 549-0671

Technology Applications Center
University of New Mexico
Albuquerque, New Mexico 87131
Stanley A. Morain, Ph.D., director
Phone: (505) 277-3622

Southern Technology Applications Center
Progress Center, Box 24
1 Progress Boulevard
Alachua, Florida 32615
J. Ronald Thornton, director
Phone: (904) 462-3913
(800) 354-4832 (FL only)
(800) 225-0308 (toll free, US)

NASA/UK Technology Applications Program
10 Kinkead Hall
University of Kentucky
Lexington, Kentucky 40506
William R. Stong, director
Phone: (606) 257-6322

NASA/SU Industrial Applications Center
Southern University
Department of Computer Science
Baton Rouge, Louisiana 70813-2065
John Hubbell, Ph.D., director
Phone: (504) 771-6272

○ Industrial Application Center Affiliates

Alabama
Dr. Bernie Schroer
Johnson Research Center
University of Alabama-Huntsville
Huntsville, Alabama 35899
Phone: (205)895-6361

Alaska
Mr. Patrick Anderson, Director
Alaska Economic Development Center
University of Alaska—Juneau
1108 F Street, Juneau, Alaska 99801
Phone: (907) 789-4402

Arkansas
Mr. Homer Presnall
Industrial Research and Extension Center
University of Arkansas
33rd and University
Little Rock, Arkansas 72204
Phone: (501) 371-1971

Dr. George Marsh
Center for Interactive Technology
Arkansas Center for Technology Transfer
Engineering Experiment Station
University of Arkansas
Fayetteville, Arkansas 72701
Phone: (501) 575-5399

Mr. Alan A. Gumbel
Arkansas Science and Technology Authority
100 Main Street, Suite 450
Little Rock, Arkansas 72201
Phone: (501) 371-3554

California
● See Industrial Applications Center, University of Southern California

Colorado
Ms. Patti Martillaro, Director
Business Advancement Center
University of Colorado
1690 38th Street, Suite 101
Boulder, Colorado 80301
Phone: (303) 444-5723

Florida
● See Southern Technology Applications Center, University of Florida

Georgia
Dr. David Swanson
Economic Development Laboratory
Georgia Institute of Technology Research Institute
Atlanta, Georgia 30332
Phone: (404) 894-3863

Hawaii
Ms. Angela Williams
Pacific Business Center Program
University of Hawaii
2404 Maile Way
Honolulu, Hawaii 96822
Phone: (808) 948-6286

Idaho
Mr. Ronald R. Hall, Director
Idaho Business Development Center
Boise State University
1910 University Drive
Boise, Idaho 83725
Phone: (208) 385-1640

Illinois
Mr. Brian Frost
c/o Technology Transfer Center
Argonne National Laboratory
Argonne, Illinois 60439
Phone: (312) 972-7694
(312) 790-0749

Indiana
● See Aerospace Research Applications Center, Indianapolis Center for Advanced Research

Iowa
Mr. Lloyd Anderson
CIRAS—Iowa State University
205 Engineering Annex
Ames, Iowa 50011
Phone: (515) 294-3420

Kansas
Mr. Kevin Carr
Kansas Technology Enterprise Corporation
112 West 6th Street, Suite 400
Topeka, Kansas 66603
Phone: (913) 296-5272

Kentucky
● See NASA/UK Technology Applications Program, University of Kentucky

Louisiana
● See NASA/SU Industrial Applications Center, Southern University

Maine
Mr. Ivan Most, P.E. Director
Production Technology Center
University of Southern Maine
37 College Avenue
Gorham, Maine 04038
Phone: (207) 780-5439

Maryland
Mr. Travis Walton
Technology Extension Service
Engineering Research Center
University of Maryland
College Park, Maryland 20742
Phone: (301) 454-1941

Minnesota
Mr. Gil Young
Governor's Office of Science and Technology
900 American Center Building
150 E. Kellogg Boulevard
St. Paul, Minnesota 55101
Phone: (612) 297-4368

Mississippi
Dr. James Miller, Director
Mississippi Technology Transfer Office
Stennis Space Center, Mississippi 39529
Phone: (601) 688-3144

Missouri
Mr. Fred Goss
Technology Search Program
304 Harris Hall
University of Missouri-Rolla
Rolla, Missouri 65401
Phone: (314) 341-4004

Montana
Dr. William R. Taylor, Director
University Technical Assistance Program
College of Engineering
402 Roberts Hall
Montana State University
Bozeman, Montana 59717-0007
Phone: (406) 994-3812

Nebraska
Mr. Thomas Spilker
Nebraska Technical Assistance Center
W191 Nebraska Hall
University of Nebraska
Lincoln, Nebraska 68588
Phone: (402) 472-5600

Nevada
Mr. Bernard "Gus" Zuzo
Desert Research Institute
P.O. Box 60220
Reno, Nevada 89506
Phone: (702) 673-7388

New England
● See Industrial Applications Center,
NERAC, Incorporated

New Jersey
Mr. Thomas Allen, Director of
Operations
Southern New Jersey Technology
Consortium
236 Doughty Road
Pleasantville, New Jersey 08232
Phone: (609) 641-1008

New Mexico
● See Industrial Application Center
TAC, University of New Mexico

Mr. James Ray
Business Assistance and Resource Center
University of New Mexico
1920 Lomas NE
Albuquerque, New Mexico 87131
Phone: (505) 277-3541

North Carolina
● See Industrial Applications Center
North Carolina Science and Technology
Research Center

North Dakota
Mr. Leonard Christianson, Director
Technical Services Center for
Innovation and Business
Development
Box 8103, University Station
Grand Fork, North Dakota 58202
Phone: (701) 777-3796

Ohio
Mr. Jeff Schick, Director
Ohio Technology Transfer Organization
65 East State Street, Suite 200
Columbus, Ohio 43266-0330
Phone: (614) 466-4286

Oklahoma
● See Central Industrial Applications
Center, Rural Enterprises, Inc.

Oregon
Mr. Terry Edvalson, Director
Regional Services Institute
Eastern Oregon State College
8th and K
La Grande, Oregon 97850
Phone: (503) 963-2171

Pennsylvania
● See Industrial Applications Center,
University of Pittsburgh

South Carolina
Dr. Ted Gasper
The State Board for Technical and
Comprehensive Education
111 Executive Center Drive, Room 103
Columbia, South Carolina 29210
Phone: (803) 737-9320

South Dakota
Mr. LaDell Swiden
College of Engineering, Box 507
South Dakota State University
Brookings, South Dakota 57007-0199
Phone: (605) 688-4184

Tennessee
Mr. T.C. Parsons, Executive Director
Center for Industrial Services
University of Tennessee
226 Capitol Boulevard Building,
Suite 401
Nashville, Tennessee 37219-1804
Phone: (615) 242-2456

Texas
Dr. Helen Dorsey, Director
Technology Business Development
Texas Engineering Experiment Station
College Station, Texas 77843-3369
Phone: (405) 845-0538

Vermont
Mr. Graeme H. Freeman
Agency of Development and
Community Affairs
State of Vermont
Montpelier, Vermont 05602
Phone: (802) 828-3221

Virginia
Dr. Gene Calvert
Center for Innovative Technology
2214 Rock Hill Road
Herndon, Virginia 22070
Phone: (703) 689-3000

Washington
Dr. Lyle M. Anderson, State Director
Small Business Development Center
Washington State University
441 Todd Hall
Pullman, Washington 99164-4740
Phone: (509) 335-1576

West Virginia
Mr. Donald F. Butcher, President
The Software Valley Association
c/o Department of Statistics and
Computer Science
311 Knapp Hall
University of West Virginia
Morgantown, West Virginia 26056
Phone: (304) 293-3607

Science and Technology Center
c/o Dr. David Powers
Vice Chancellor for Academic Affairs
Board of Regents
950 Kanawha Boulevard, East
Charleston, West Virginia 25301
Phone: (304) 348-2101

Wyoming
Ms. Gail Mathews
Institute of Business Management
Services
College of Commerce and Industry
University of Wyoming
P.O. Box 3275, University Station
Laramie, Wyoming 82071
Phone: (307) 766-2363

■ Computer Software Management and Information Center

COSMIC
382 E. Broad Street
University of Georgia
Athens, Georgia 30602
John A. Gibson, director
Phone: (404) 542-3265

▲ Application Teams

Research Triangle Institute
Post Office Box 12194
Research Triangle Park,
North Carolina 27709
Doris Rouse, Ph.D., director
Phone: (919) 541-6980

Rural Enterprises, Inc.
P.O. Box 1335
Durant, Oklahoma 74702
Steve R. Hardy, president
Phone: (405) 924-5094

NASA Scientific and Technical Information Facility

Technology Utilization Office
P.O. Box 8757
Baltimore, Maryland 21240
Walter Heiland, manager
Phone: (301) 859-5300 extension 241

Spinoff Team:
Project Manager
Walter Heiland

Technology Associate:
Linda Watts

Technology Associate:
Jane Lynn-Jones

Graphic Services:
Justin Associates

Photography:
Kevin Wilson